COMPACT LOW-VOLTAGE AND HIGH-SPEED CMOS, BiCMOS AND BIPOLAR OPERATIONAL AMPLIFIERS

THE KLUWER INTERNATIONAL SERIES IN ENGINEERING AND COMPUTER SCIENCE

ANALOG CIRCUITS AND SIGNAL PROCESSING

Consulting Editor: **Mohammed Ismail**. *Ohio State University*

Related Titles:

NEUROMORPHIC SYSTEMS ENGINEERING: ***Neural Networks in Silicon***, *edited by Tor Sverre Lande*; ISBN: 0-7923-8158-0

DESIGN OF MODULATORS FOR OVERSAMPLED CONVERTERS, *Feng Wang, Ramesh Harjani*, ISBN: 0-7923-8063-0

SYMBOLIC ANALYSIS IN ANALOG INTEGRATED CIRCUIT DESIGN, *Henrik Floberg*, ISBN: 0-7923-9969-2

SWITCHED-CURRENT DESIGN AND IMPLEMENTATION OF OVERSAMPLING A/D CONVERTERS, *Nianxiong Tan*, ISBN: 0-7923-9963-3

CMOS WIRELESS TRANSCEIVER DESIGN, *Jan Crols, Michiel Steyaert*, ISBN: 0-7923-9960-9

DESIGN OF LOW-VOLTAGE, LOW-POWER OPERATIONAL AMPLIFIER CELLS, *Ron Hogervorst, Johan H. Huijsing*, ISBN: 0-7923-9781-9

VLSI-COMPATIBLE IMPLEMENTATIONS FOR ARTIFICIAL NEURAL NETWORKS, *Sied Mehdi Fakhraie, Kenneth Carless Smith*, ISBN: 0-7923-9825-4

CHARACTERIZATION METHODS FOR SUBMICRON MOSFETs, edited by *Hisham Haddara*, ISBN: 0-7923-9695-2

LOW-VOLTAGE LOW-POWER ANALOG INTEGRATED CIRCUITS, edited by *Wouter Serdijn*, ISBN: 0-7923-9608-1

INTEGRATED VIDEO-FREQUENCY CONTINUOUS-TIME FILTERS: ***High-Performance Realizations in BiCMOS***, *Scott D. Willingham, Ken Martin*, ISBN: 0-7923-9595-6

FEED-FORWARD NEURAL NETWORKS: ***Vector Decomposition Analysis, Modelling and Analog Implementation,*** *Anne-Johan Annema*, ISBN: 0-7923-9567-0

FREQUENCY COMPENSATION TECHNIQUES LOW-POWER OPERATIONAL AMPLIFIERS, *Ruud Easchauzier, Johan Huijsing*, ISBN: 0-7923-9565-4

ANALOG SIGNAL GENERATION FOR BIST OF MIXED-SIGNAL INTEGRATED CIRCUITS, *Gordon W. Roberts, Albert K. Lu*, ISBN: 0-7923-9564-6

INTEGRATED FIBER-OPTIC RECEIVERS, *Aaron Buchwald, Kenneth W. Martin*, ISBN: 0-7923-9549-2

MODELING WITH AN ANALOG HARDWARE DESCRIPTION LANGUAGE, *H. Alan Mantooth, Mike Fiegenbaum*, ISBN: 0-7923-9516-6

LOW-VOLTAGE CMOS OPERATIONAL AMPLIFIERS: ***Theory, Design and Implementation,*** *Satoshi Sakurai, Mohammed Ismail*, ISBN: 0-7923-9507-7

ANALYSIS AND SYNTHESIS OF MOS TRANSLINEAR CIRCUITS, *Remco J. Wiegerink*, ISBN: 0-7923-9390-2

COMPUTER-AIDED DESIGN OF ANALOG CIRCUITS AND SYSTEMS, *L. Richard Carley*, Ronald S. Gyurcsik, ISBN: 0-7923-9351-1

HIGH-PERFORMANCE CMOS CONTINUOUS-TIME FILTERS, *José Silva-Martínez, Michiel Steyaert, Willy Sansen*, ISBN: 0-7923-9339-2

SYMBOLIC ANALYSIS OF ANALOG CIRCUITS: ***Techniques and Applications***, *Lawrence P. Huelsman, Georges G. E. Gielen*, ISBN: 0-7923-9324-4

DESIGN OF LOW-VOLTAGE BIPOLAR OPERATIONAL AMPLIFIERS, *M. Jeroen Fonderie, Johan H. Huijsing*, ISBN: 0-7923-9317-1

STATISTICAL MODELING FOR COMPUTER-AIDED DESIGN OF MOS VLSI CIRCUITS, *Christopher Michael, Mohammed Ismail*, ISBN: 0-7923-9299-X

SELECTIVE LINEAR-PHASE SWITCHED-CAPACITOR AND DIGITAL FILTERS, *Hussein Baher*, ISBN: 0-7923-9298-1

ANALOG CMOS FILTERS FOR VERY HIGH FREQUENCIES, *Bram Nauta*, ISBN: 0-7923-9272-8

ANALOG VLSI NEURAL NETWORKS, *Yoshiyasu Takefuji*, ISBN: 0-7923-9273-6

COMPACT LOW-VOLTAGE AND HIGH-SPEED CMOS, BiCMOS AND BIPOLAR OPERATIONAL AMPLIFIERS

by

Klaas-Jan de Langen
Philips Semiconductors, Inc.,
Sunnyvale, California

and

Johan H. Huijsing
Delft University of Technology,
The Netherlands

SPRINGER SCIENCE+BUSINESS MEDIA, LLC

A C.I.P. Catalogue record for this book is available from the Library of Congress.

ISBN 978-1-4419-5102-1 **ISBN 978-1-4757-2993-1 (eBook)**
DOI 10.1007/978-1-4757-2993-1

This printing is a digital duplication of the original edition.

Printed on acid-free paper

Second Printing 2003

Originally published by Kluwer Academic Publishers in 1999
Softcover reprint of the hardcover 1st edition 1999

Table of Contents

Preface

"Compact Low-Voltage and High-Speed CMOS, BiCMOS and Bipolar Operational Amplifiers" focuses on the design of integrated operational amplifiers that approach the limits of low supply voltage or very high bandwidth. The resulting realizations span the whole field of applications from micro-power CMOS VLSI amplifiers to 1-GHz bipolar amplifiers. The work is especially useful for professional designers who like to improve their amplifier designs to match the state of the art. The book is also of interest to graduate students in the field of analog circuit design.

After discussing the fundamentals of low-voltage power-efficient operational amplifier design in the first chapter, chapter 2 discusses input stages. First, basic input stages are discussed, followed by rail-to-rail input stages with constant transconductance for constant bandwidth. New input-stage circuits that result in compact solutions are presented (section 2.4). Finally, new compact PTAT bias-current generators are discussed by which a constant transconductance of CMOS input stages as a function of temperature can be obtained (section 2.5).

In chapter 3, output stages are discussed. First, rail-to-rail output stages are treated. A rail-to-rail output-voltage swing can be obtained by using a common-source connected or common-emitter connected output stage. Class-AB biasing assures a compromise between power consumption and distortion. Two types of class-AB biasing principles can be distinguished: feedforward class-AB control and feedback class-AB control. Numerous examples of both principles are given in CMOS as well as in bipolar technology. In addition, the stability of the class-AB control loop of feedback biased output stages is analyzed (section 3.5). Especially interesting is the feedback class-AB biasing with a simple minimum selector which can be applied in CMOS as well as in bipolar technology (section 3.5). Secondly, all-NPN output stages are presented. Thanks to the excellent performance of NPN transistors, all-NPN output stages are still important for achieving very high bandwidth or very high output-current capability. Finally, an overview of output stages in BiCMOS technology is given (section 3.7). For bipolar output stages in bipolar and BiCMOS amplifiers very important protection circuits are discussed which limit the output current and the saturation of output transistors (section 3.8).

Chapter 4 treats the overall design of operational amplifiers. In order to obtain sufficient gain, an operational amplifier consists of several cascaded gain stages. Since each stage introduces a dominant pole, frequency compensation is required to shape the frequency response. The only frequency compensation techniques suited to general purpose amplifiers are based on Miller compensation. Using Miller compensation the bandwidth reduces when the number of stages increases. The optimization of two-stage Miller compensated amplifiers is discussed. If the supply voltage is equal to one gate-source voltage and one saturation, at least three stages are needed to achieve sufficient gain in bipolar technology, in BiCMOS technology, and in CMOS technology. A three-stage amplifier can be compensated using nested-Miller compensation. Two new active versions of nested-Miller compensation are introduced: cascoded-nested-Miller compensation (section 4.6.2) and mirrored-nested-Miller compensation (section 4.6.3).

In the final chapter, chapter 5, many realizations are presented. Two amplifiers have been designed in bipolar technology. The first amplifier is a wideband amplifier with a very high unity-gain frequency of 1 *GHz*. The second amplifier has a rail-to-rail class-AB output stage with a high ratio between the maximum output current and the quiescent current. Five amplifiers have been designed in CMOS technology. Three Amplifiers are based on a compact topology that can operate on a supply voltages down to one gate-source voltage and two saturation voltages which amounts to about 1.4 *V*. The two other CMOS amplifiers are based on an ultimate-low-voltage topology and can operate on supply voltages down to one gate-source voltage and one saturation voltages which amounts to about 1.2 *V*. In BiCMOS technology, five amplifiers have been designed. The first two amplifiers are based on a compact topology. Two other amplifiers are designed to operate on low supply voltages down to 1.3 *V*. The final amplifier uses a powerful combination of an all-NPN output stage driven by a PMOS intermediate stage to obtain a power-efficient wideband amplifier. The amplifier has a unity-gain frequency of 200 *MHz* and can operate down to 2.5 *V*.

Klaas-Jan de Langen
Johan H. Huijsing

Delft, April 6, 1999.

List of symbols

Symbol	Quantity	Unit
β	transconductance factor	A/V^2
β_F	forward current gain of bipolar transistor	-
β_R	reverse current gain of bipolar transistor	-
γ	bulk-threshold parameter	$1/V^{1/2}$
μ	voltage gain	-
μ_n	mobility of electrons	cm^2/Vs
μ_n	mobility of holes	cm^2/Vs
σ	real part of the complex frequency	rad/s
τ	time constant	s
ϕ_f	flat-band voltage	V
φ_m	phase margin	°
ω	frequency	rad/s
ω_d	doublet frequency	rad/s
ω_d	unity-gain frequency of uncompensated system	rad/s
ω_p	pole frequency	rad/s
ω_u	unity-gain frequency	rad/s
ω_z	zero frequency	rad/s
A_0	dc open-loop gain	dB
B_e	effective bandwidth	Hz
CMRR	common-mode rejection ratio	dB
C	symbol for capacitor	F
C_{BE}	base-emitter capacitance	F
C_{GS}	gate-source capacitance	F
C_M	Miller capacitor	F
C_L	load capacitor	F
C_{ox}	normalized oxide capacitance	F/m^2
DR_{max}	maximum dynamic range	dB
f	frequency	Hz
f_T	transit frequency	Hz
g_m	transconductance	Ω^{-1}
i	small-signal current	A
i_{in}	small-signal input current	A
I	current	A

I_B	base current of bipolar transistor	A
I_{BIAS}	bias current	A
I_C	collector current of bipolar transistor	A
I_D	drain current of MOS transistor	A
I_M	minimum current	A
I_Q	quiescent current	A
I_{REF}	reference current	A
I_S	saturation current of bipolar transistor	A
$j\omega$	imaginary part of the complex frequency	rad/s
k	Boltzmann's constant, $1.3805 \cdot 10^{-23}$	J/K
K_n	flicker noise component of NMOS transistor	V^2F
K_p	flicker noise component of PMOS transistor	V^2F
L	channel length of MOS transistor	m
M	symbol for MOS transistor	-
M	number of cascode stages	-
n	weak inversion slope factor of MOS transistor	-
n	scaling factor	-
N	number of stages	-
P	power	W
P_{sup}	supply power	W
q	electron charge, $1.6 \cdot 10^{-19}$	C
Q	symbol for bipolar transistor	-
r_{be}	base-emitter resistance of bipolar transistor	Ω
r_{ds}	drain-source resistance of MOS transistor	Ω
r_o	output resistance of bipolar transistor	Ω
R	symbol for resistor	-
R_B	base bulk resistor of bipolar transistor	Ω
R_C	collector bulk resistor of bipolar transistor	Ω
R_L	load resistor	Ω
S_d	doublet spacing	-
T	absolute temperature	K
T_S	settling time	s
v	small-signal voltage	V
v_{in}	small-signal input voltage	V
v_{out}	small-signal output voltage	V
v_{th}^2	squared equivalent thermal input noise voltage	V^2/Hz
V_{ss}	signal voltage with maximum amplitude	V
V_{BE}	base-emitter voltage of bipolar transistor	V
V_{CC}	positive supply voltage of bipolar circuits	V
V_{CE}	collector-emitter voltage of bipolar transistor	V
V_{CM}	common-mode input voltage	V

V_{DD}	positive supply voltage of MOS circuits	V
V_{DS}	drain-source voltage of MOS transistor	V
V_{EE}	negative supply voltage of bipolar circuits	V
V_{GS}	gate-source voltage of MOS transistor	V
V_{IN}	input voltage	V
V_N	noise voltage	V
V_{NN}	negative supply voltage of BiCMOS circuits	V
V_{OS}	offset voltage	V
V_{OUT}	output voltage	V
V_{PP}	positive supply voltage of BiCMOS circuits	V
V_{Sat}	saturation voltage	V
V_{SB}	source-bulk voltage of MOS transistors	V
V_{SS}	negative supply voltage of MOS circuits	V
V_{SUP}	supply voltage	V
$V_{SUP,min}$	minimum supply voltage	V
V_T	thermal voltage	V
V_{Th}	threshold voltage	V
V_{T0}	threshold voltage at V_{SB}=0V	V
W	channel width of MOS transistor	m
Z	impedance	Ω

Introduction 1

Bandwidth, power and supply voltage are the main parameters in modern electronic design. The demand for higher bandwidth, smaller supply voltage and lower power consumption is driven by the increasing complexity of electronic equipment. To be able to integrate more functions and more complex functions on a single integrated circuit, the minimum feature size continues to reduce, forcing smaller supply voltages and lower power consumption. The impact on the design of operational amplifiers has been reported in previous work on bipolar amplifiers [1] and CMOS amplifiers [2]. Further, research has been done on the frequency limitations of amplifiers [3]. Now, the availability of more complicated IC processes such as BiCMOS technology opens up new opportunities. The advantages of BiCMOS technology over CMOS circuits have been shown for digital circuits [4,5], but utilizing the BiCMOS process in amplifier design has only briefly been attempted [6]. Therefore, exploiting the full possibilities of the BiCMOS technology is the main theme of this work. Using BiCMOS technology, we have shown that the advantages of bipolar transistors and the advantages of CMOS transistors can be combined, yielding smaller die area or an improved trade-off between bandwidth and power consumption. This research also produced advances in bipolar amplifier design and in CMOS VLSI amplifier design. Optimization of bandwidth using bipolar technology resulted in a 1-GHz operational amplifier, while the allowable supply voltage of compact CMOS amplifiers has been further reduced, and the topology of ultimate-low-voltage CMOS amplifiers has been further condensed.

In this chapter we briefly review the impact of low-supply voltages and low-power consumption on the signal handling of analog circuits. Further, the limitations of transistors with regard to low-voltage design are discussed and, finally, the consequences for the design of amplifiers are presented.

1.1 Low-voltage limitations

The main concern in low-voltage circuits is to achieve sufficient dynamic range. The dynamic range of low-voltage low-power amplifiers is limited on the upper side by the low-supply power, and on the lower side by the noise power in resistors and in active devices. It can be calculated that the maximum dynamic range for a simple example, where a signal-processing resistor R_1 is driven by an amplifier, is given by [7]

$$DR_{max} = \frac{V_{SS}^2}{V_N^2} = \frac{\pi}{16kTB_e} P_{sup} \tag{1-1}$$

in which V_{SS} is the root-mean-square value of a sine-wave with maximum amplitude, V_N the noise voltage, k Boltzmann's constant, T the absolute temperature, B_e the effective bandwidth, and P_{sup} the supply power. The resistor is assumed to be the only noise source in the circuit while the amplifier is biased in class B.

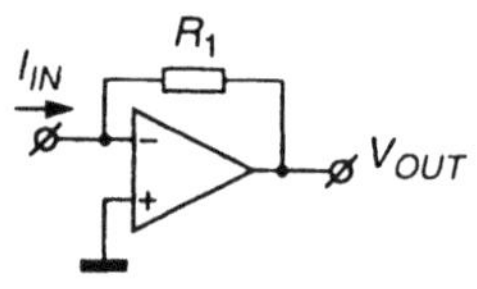

Fig. 1-1 *Current-to-voltage converter.*

A simple example is shown in the current-to-voltage converter shown in Fig. 1-1. Using a resistor R_1 of 10 $k\Omega$, a supply voltage of 2 V, and a bandwidth of 1 MHz at a temperature of 300 K, the maximum dynamic range DR_{max} is 95 dB at a power consumption of 64 μW. This maximum can only be obtained if the signal-processing resistor can be driven from rail to rail and in class-B. If the supply voltage is reduced by a factor of two to 1 V, DR_{max} is only 89 dB at a power consumption of 16 μW. The reduction in dynamic range can be counteracted by reducing the value of the resistor by a factor of four, so that the power consumption remains the same. In that case a higher output current capability of the output stage of the amplifier is required. Another possibility is to reduce the bandwidth by a factor of four. This can be obtained by increasing the compensation capacitors of the amplifier by a factor of four which consumes a lot of die area.

The dynamic range is further influenced by the implementation of the amplifier. If, for example, resistor R_l is driven in class-A instead of class-B, the dynamic range will be minimally a factor π, or 5 dB, lower. This loss can be much higher, especially in situations where the signals are much smaller than their maximum values most of the time, such as in circuits that process speech, music etc. If the signals are limited to 1/3 of the supply voltage, for example when a diode voltage V_{BE} is lost at a supply voltage of 1 V, the dynamic range loses another factor 3, or 5 dB. Concluding, amplifiers should employ class-B stages and should be able to handle signals from rail to rail.

Another problem is that, depending on how low the actual supply voltage is, several topologies cannot be used. As long as the supply voltage is larger than two base-emitter voltages and a saturation voltage that is sufficient to bias a current source, a Darlington configuration can still be used. If the supply voltage is reduced a Darlington cannot be used. Cascodes and cascoded current sources can only be used in CMOS technology if the supply voltage is larger than a gate-source voltage and two saturation voltages. If the supply voltage is further reduced the ultimate supply-voltage limit is reached at one gate-source voltage and one saturation voltage. Under these circumstances only a cascade of common-source stages can be applied as shown in Fig. 1-2, while the application of folded cascodes and differential pairs is limited to special cases. The minimum supply voltage of this circuit is the lowest possible supply voltage and is given by

$$V_{SUP,min} = V_{GS} + V_{Sat} \tag{1-2}$$

where V_{GS} is the gate-source voltage and V_{Sat} is the saturation voltage which is necessary for proper operation of the current sources. In bipolar technology, the minimum supply voltage of a cascade of common-emitter stages is given by

$$V_{SUP,min} = V_{BE} + V_{Sat} \tag{1-3}$$

where V_{BE} is the base-emitter voltage. To compensate a multi-stage amplifier as shown in Fig. 1-2, frequency compensation techniques are needed that can handle the multi-stage topology. In order to minimize the power consumption of analog circuits the frequency compensation techniques should be efficient so that the minimum required supply current is obtained for a given bandwidth [3]. The basic voltage limitations and bandwidth limitations of transistors are discussed in the next section.

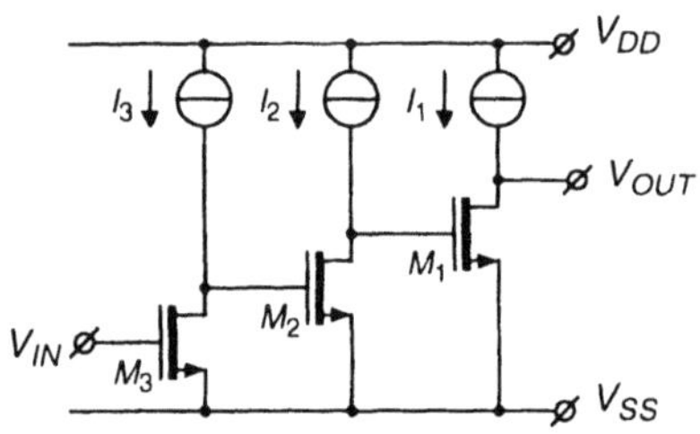

Fig. 1-2 *Basic topology of ultra-low-voltage amplifier consisting of a cascade of common-source stages.*

1.2 Transistor limitations

In this section we briefly discuss the main transistor parameters that determine the behavior of low-voltage circuits. The biasing voltages which determine the minimum supply voltage are discussed. Further, gain and bandwidth are important figures that are considered. First, bipolar transistors are discussed then CMOS transistors are treated.

1.2.1 Bipolar junction transistor

The relation between base-emitter voltage V_{BE} and collector current I_C is described by the well-known expression

$$V_{BE} = \frac{kT}{q} \ln\left(\frac{I_C}{I_S}\right) \tag{1-4}$$

where q is the electron charge and I_S the saturation current. Because of the exponential characteristic, the base-emitter voltage only depends weakly on the collector current so that large currents can be handled at low supply voltages. The base-emitter voltage is of the order of 600 to 800 mV at room temperature, depending on the process and current density. The minimum collector-emitter voltage or saturation voltage V_{Sat} is limited by forward biasing of the base-collector junction. The saturation voltage V_{Sat} can be calculated using the Ebers-Moll model. However, it should be noted that most of the base-current flows via a parasitic transistor to substrate. An example of an NPN transistor in IC technology is shown in Fig. 1-3. We see that the transistor together with the P-type substrate forms a stack of NPNP layers which can be represented by the circuit model shown on the right-hand side. The base-collector junction forms the base-

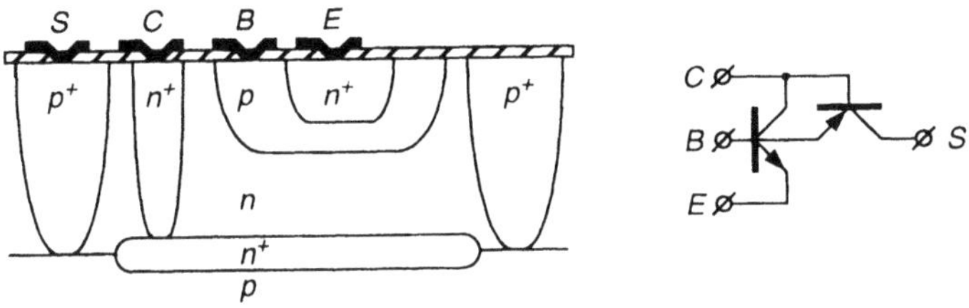

Fig. 1-3 *Cross section of NPN transistor and equivalent circuit.*

emitter junction of a PNP transistor which has a collector connected to substrate. Therefore, if the NPN transistor saturates, its base-emitter junction is forward biased and the PNP transistor is turned on. Thus, most of the base current of the PNP transistor is injected into the substrate. Assuming most of the base current flows into the substrate, which means that the current gain of the substrate PNP is considerably larger than unity, we find

$$V_{Sat} = I_C R_C + \frac{kT}{q} \ln\left(\frac{\frac{1}{\beta_R}\frac{I_C}{I_B}}{1 - \frac{1}{\beta_F}\frac{I_C}{I_B}} \right) \qquad (1\text{-}5)$$

where R_C is the collector bulk resistance, I_B the base current, β_F the forward current gain, and β_R the reverse current gain. The voltage V_{Sat} mainly depends on the collector current, the collector bulk resistance, the temperature, and the available base current. Under normal circumstances the saturation voltage is of the order of 100 *mV*. Substituting the above-found values in equation (1-3), we find that operation of bipolar circuits is possible down to a supply voltage of 1 *V*.

We have found the bias voltage limitations and it is now time to consider the signal behavior. The main signal parameters of interest for amplifier design are gain and bandwidth. The maximum voltage gain of bipolar transistors is relatively high. However, due to the low input impedance, the gain of bipolar stages in a cascade is limited to the current gain β_F. The bandwidth of the bipolar transistor is determined by the transit frequency f_T which is in first order given by

$$f_T = \frac{g_m}{C_{be}} = \frac{q}{kTC_{be}} I_C \qquad (1\text{-}6)$$

1.3 Low-voltage power-efficient operational amplifier design

The required efficiency at low-voltage and low-power conditions necessitates rail-to-rail signal handling at the input and the output of the amplifier. Furthermore, efficient overall topologies are needed for combining sufficient gain with high bandwidth. So, we briefly introduce the concepts of rail-to-rail input stages, rail-to-rail output stages and efficient overall topologies in this section.

In order to process input signals with the maximum signal voltage at a certain supply voltage, rail-to-rail input stages are needed. An application example is an operational amplifier connected as a rail-to-rail voltage follower. But even if we do not actually need the full rail-to-rail range, it may often be preferred to process input voltages close to either the ground rail, or the positive supply rail, without using level-shift networks.

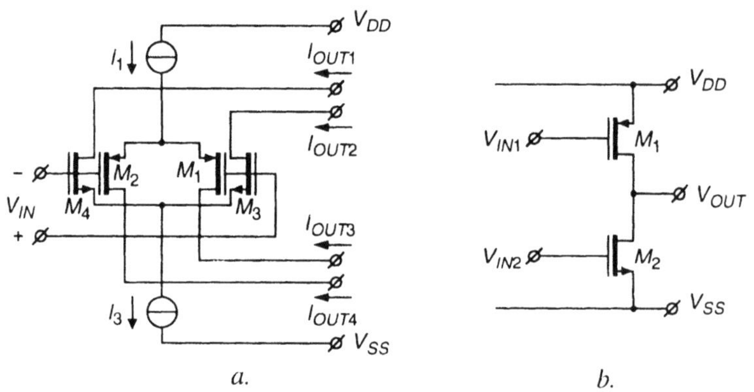

Fig. 1-4 *Rail-to-rail amplifier stages: rail-to-rail input stage (a) and rail-to-rail output stage (b).*

In order to obtain a rail-to-rail input stage, P-type transistors are needed to reach the negative supply rail, while N-type transistors should be applied to reach the positive supply rail. An example of a rail-to-rail input stage is depicted in Fig. 1-4a. The circuit consists of a PMOS differential pair M_1, M_2 in parallel with an NMOS differential pair M_3, M_4. For high common-mode voltages the NMOS differential pair can operate, while for low common-mode voltages the PMOS differential pair is able to function. A full rail-to-rail range can then be achieved by properly combining the outputs of the two input stages. A further requirement is that the transconductance of the rail-to-rail input stage must be largely constant as a function of the common-mode input voltage for reasons of optimal frequency compensation and distortion. Consequently, a control circuit is needed that regulates the transconductance of the input stage [2]. The implementation of the g_m-control circuit is discussed in chapter 3.

In order to process output signals with the maximum signal voltage at a certain supply voltage, output stages should have an output voltage range from rail to rail, so that the supply voltage is used efficiently. Allowing rail-to-rail output swing is possible using common-source connected transistors, as demonstrated in the output stage shown in Fig. 1-4b. The stage consists of a PMOS transistor M_1 and an NMOS transistor M_2 which are both connected with their source terminal to the supply rails. In this way the output voltage can approach both supply rails within a saturation voltage. Further, the supply current should be used efficiently by biasing the output-stage in class AB. This is extensively treated in chapter 3.

A final step in low-voltage low-power amplifier design is to combine sufficient gain with high bandwidth while keeping the power consumption as low as possible. A number of stages are needed to implement the required gain. Depending on the process and the available supply voltage different strategies are found. In CMOS technology, two stages can be sufficient if the supply voltage is large enough so that cascoding techniques can be employed to improve the gain of single stages. In bipolar technology, or in CMOS technology at the lowest supply voltage, at least three stages are required. To compensate such a multi-stage amplifier, efficient frequency compensation techniques should be applied so that the gain can be combined with high bandwidth for the consumed supply current. These frequency compensation techniques are considered in chapter 4.

In BiCMOS technology, high efficiency can be obtained by employing the advantages of bipolar and CMOS transistors. The advantage of the bipolar transistor is its high transconductance which leads to high bandwidth and low offset and noise voltage. A disadvantage, however, is the base current which must be delivered by the driving circuit and which limits the achievable gain. The advantage of MOS transistors is their very high input impedance which enables high gain with compact topologies using cascoding. Further, no input current has to be supplied by a driving circuit. On the other hand, the transconductance is considerably smaller than the transconductance of bipolar transistors yielding moderate bandwidths. By combining the advantages of CMOS and bipolar transistors high gain and high bandwidth can be combined. A simplified example is depicted in Fig. 1-5. Bipolar transistors are applied where high transconductance is beneficial, while CMOS transistors are used where a high input impedance is important. A bipolar transistor Q_2 is used in the input stage where the high transconductance results in low offset voltage. In the output stage, bipolar transistor Q_1 is used because its high transconductance yields high bandwidth for a given capacitive load and current consumption. Further, the exponential relation between base-emitter voltage and collector current enables large output currents at low supply voltages. The output transistor is driven by CMOS Darlington transistor M_{11} so that the input stage does not

Input Stages 2

2.1 Introduction

The task of the input stage is to process the differential signals present at the inputs of the operational amplifier while keeping the amount of interference that is added by the input stage as small as possible. Interference includes offset, noise, and distortion. Offset is caused by a mismatch of devices, while noise is generated by active components and resistors. Interference is also generated by cross-talk from common-mode input signals due to a finite common-mode rejection. If the amplifier is used in applications where the common-mode voltage varies, as a non-inverting amplifier for example, the finite common-mode rejection causes distortion. Due to the low-supply voltages used in modern electronics, the input-voltage range of a conventional input stage is very small. Especially if the applications of the operational amplifier include a buffer connected configuration, the common-mode input-voltage range should be as large as possible so that an input stage with rail-to-rail input-voltage range is needed. Basic input stages, input stages with rail-to-rail input range, and the biasing required by input stages are discussed in this chapter.

In section 2.2, basic input-stage configurations in CMOS and bipolar technology are discussed along with their most important performance parameters such as offset, noise, and common-mode rejection. Thereafter, the common-mode input range is maximized using rail-to-rail input stages in section 2.3. Rail-to-rail input stages with constant common-mode output currents are discussed in section 2.4. Section 3.5 presents bias circuits for making the transconductance of the input stages temperature independent and, finally, conclusions are drawn in section 2.6.

2.2 Basic input stages

In this section the main parameters of basic CMOS and bipolar input stages are discussed. A conventional input stage in CMOS technology is depicted in Fig. 2-1. The input stage consists of a differential input stage M_1, M_2, folded cascode M_{41}, M_{42}, and PMOS current mirror M_{31}, M_{32} that sums the differential output signals of the differential stage yielding a single-ended output. A folded cascode is used to make the input voltage range as large as possible. The input voltage range is limited on the upper

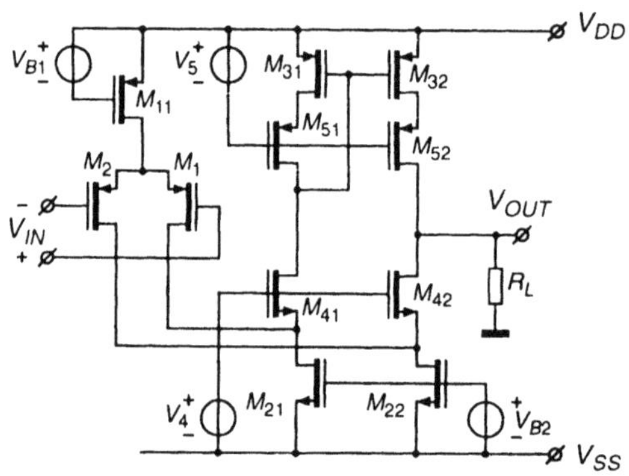

Fig. 2-1 *Basic input stage in CMOS technology consisting of PMOS differential pair followed by an NMOS folded cascode summing circuit.*

side by the positive supply rail V_{DD} minus a saturation voltage and a gate-source voltage for proper operation of the differential input stage M_1, M_2, and its current source M_{11}. The input-voltage range is limited on the lower side by the negative supply voltage V_{SS}, minus the gate-source voltage of the input transistors plus two saturation voltages for proper operation of current sources M_{21} and M_{22} and the input-stage transistors. If V_{SS} is connected to ground and the supply voltage V_{DD}-V_{SS} is 1.5 V, the input range extends from -0.5V to +0.3V which is quite small.

The offset voltage of the input stage is mainly determined by the mismatch of threshold voltages and the mismatch of transconductance factors

$$\beta = \frac{1}{2}\mu C_{ox}\frac{W}{L} \tag{2-1}$$

The offset voltage V_{OS} is given by

$$V_{OS} = \Delta V_{Th1} + \frac{g_{m2}}{g_{m1}}\Delta V_{Th2} + \frac{g_{m3}}{g_{m1}}\Delta V_{Th3} + \frac{I}{g_{m1}}\left(\frac{\Delta\beta_1}{\beta_1} + 2\frac{\Delta\beta_2}{\beta_2} + \frac{\Delta\beta_3}{\beta_3}\right) \tag{2-2}$$

where ΔV_{Th1} is the mismatch between the threshold voltages of M_1 and M_2, ΔV_{Th2} the mismatch between the threshold voltages of M_{21} and M_{22}, ΔV_{Th3} the mismatch between the threshold voltages of M_{31} and M_{32}, g_{m1} the transconductance of transistors M_1 and M_2, g_{m2} the transconductance of transistors M_{21} and M_{22}, g_{m3} the transconductance of transistors M_{31} and M_{32}, $\Delta\beta_1$ the mismatch of transconductance factors of transistors M_1 and M_2, $\Delta\beta_2$ the mismatch of transconductance factors of transistors M_{21} and M_{22}, and $\Delta\beta_3$ the mismatch of transconductance factors of transistors M_{31} and M_{32}. The offset can be minimized by making the transconductance of the current sources and the current mirror small, compared to the transconductance of the input-stage transistors, and by giving the input stage transistors a large transconductance. Since the mismatch of the threshold voltages decreases with the square root of the transistor area [1], the offset can be further reduced by giving the transistors a large area. The offset of CMOS input stages is of the order of 5 to 10 *mV*.

The common-mode rejection ratio (CMRR) is expressed by

$$CMRR = \frac{1}{\frac{1}{2g_{m1}r_{o11}}\frac{\Delta g_{m1}}{g_{m1}} + \frac{\Delta\mu_1}{\mu_1}} \tag{2-3}$$

where r_{o11} is the output impedance of current source M_{11}, Δg_{m1} the mismatch between the transconductance of M_1 and M_2, μ_1 the voltage gain of the input-stage transistors M_1 and M_2, and $\Delta\mu_1$ the mismatch between the voltage gain of M_1 and M_2. The CMRR depends on the matching of transconductance g_{m1} and the voltage gain μ_1. By increasing the output impedance r_{o11} of the tail current source, the term that depends on the transconductance can be made negligible so that the common-mode rejection is mainly determined by the mismatch of voltage gain and can be of the order of 80 *dB*.

The input-referred thermal noise of the input stage is given by

$$\overline{v_{th}^2} = \frac{16}{3}kT\frac{1}{g_{m1}}\left(1 + \sqrt{2\frac{\mu_n W_3 L_1}{\mu_p L_3 W_1}} + \sqrt{\frac{W_2 L_1}{L_2 W_1}}\right)\Delta f \tag{2-4}$$

where μ_n is the mobility of electrons, μ_p the mobility of holes, W_1/L_1 the width over length ratio of transistors M_1 and M_2, W_2/L_2 the width over length ratio of transistors M_{21} and M_{22}, W_3/L_3 the width over length ratio of transistors M_{31} and M_{32}. The thermal noise can be reduced by making the W/L ratio of the input transistors much larger than the W/L ratio of the current sources M_{21}, M_{22} and the current mirror M_{31}, M_{32}. The input-referred flicker noise voltage is expressed by

$$\overline{v_{1/f}^2} = \frac{2K_p}{W_1 L_1 C_{ox}}\left(1 + 2\frac{\mu_n K_n L_1^2}{\mu_p K_p L_2^2} + \frac{L_1^2}{L_3^2}\right)\frac{\Delta f}{f} \tag{2-5}$$

where K_p is the flicker noise coefficient of PMOS transistors, K_n the flicker noise coefficient of NMOS transistors, Δf the bandwidth of system, and f the frequency. Thus, the flicker noise can be reduced by increasing the area of the input stage transistors. Further, the flicker-noise contribution of the current sources and the mirror can be reduced by increasing the length of the transistors.

In bipolar technology, the input stage with cascoding and summing circuit is usually implemented as shown in Fig. 2-2. The circuit consists of PNP input stage Q_1, Q_2, folded

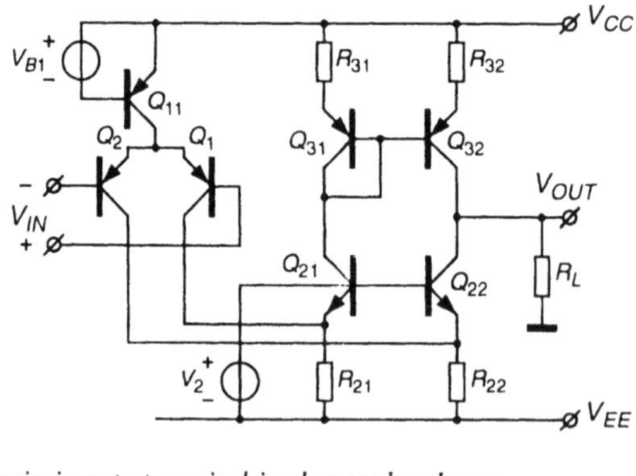

Fig. 2-2 *Basic input stage in bipolar technology.*

cascodes Q_{21}, Q_{22}, and PNP current mirror Q_{31}, Q_{32}. Emitter degeneration resistors R_{21}, R_{22}, R_{31}, R_{32} are applied to reduce the contribution of the cascodes and the mirror to the offset and noise. The common-mode input-voltage range V_{CMin} is similar to the CMOS input stage and is given by

$$V_{EE} - V_{be1} + V_{Sat1} + V_{R2} < V_{CMin} < V_{CC} - V_{Sat11} - V_{be1} \tag{2-6}$$

where V_{be1} is the base-emitter voltage of the input-stage transistors Q_1 and Q_2, V_{Sat1} the saturation voltage of transistors Q_1 and Q_2, V_{R2} the voltage across resistors R_{21} and R_{22}, and V_{Sat11} the saturation voltage of current source Q_{11}. If the supply voltage is $1.5V$ and V_{EE} is connected to ground, the input voltage range extends from $-0.5V$ to $+0.7V$ which is slightly larger than the common-mode range of the CMOS stage, but still quite limited.

The offset voltage of the bipolar input stage is given by

$$V_{OS} = \frac{kT}{q}\left[\frac{\Delta I_{S1}}{I_{S1}} + 2\frac{\Delta R_2}{R_2} + \frac{kT/q}{IR_2}\frac{\Delta I_{S2}}{I_{S2}} + \frac{1+\frac{kT/q}{IR_2}}{1+\frac{IR_3}{kT/q}}\left(\frac{IR_3}{kT/q}\frac{\Delta R_3}{R_3} + \frac{\Delta I_{S3}}{I_{S3}}\right)\right] \tag{2-7}$$

where I_{S1} is the saturation current of Q_1 and Q_2, I_{S2} the saturation current of Q_{21} and Q_{22}, I_{S3} the saturation current of Q_{31} and Q_{32}, ΔI_{S1} the mismatch between the saturation current of Q_1 and Q_2, ΔI_{S2} the mismatch between the saturation current of Q_{21} and Q_{22}, ΔI_{S3} the mismatch between the saturation current of Q_{31} and Q_{32}, R_2 the value of resistors R_{21} and R_{22}, R_3 the value of resistors R_{31} and R_{32}, ΔR_2 the mismatch between resistors R_{21} and R_{22}, and ΔR_3 the mismatch between resistors R_{31} and R_{32}. By increasing the emitter degeneration, the noise contribution of the folded-cascode transistors and the current-mirror transistors can be made small, leaving only the offset caused by the input-stage transistors and the resistors. The offset of the bipolar stage can be smaller than $1\ mV$.

The CMRR ratio is given by the same equation as the CMRR of the CMOS stage expressed by equation (2-3). Using a bipolar input stage, a CMRR of $90\ dB$ can easily be achieved.

The emitter degeneration also helps to reduce the noise. If sufficient emitter degeneration is applied, the thermal noise of the bipolar input stage is given by

$$\overline{v_{th}^2} = 4kT\left(\frac{1}{g_{m1}} + 2R_{B1} + \frac{2}{g_{m1}^2 R_2} + \frac{2}{g_{m1}^2 R_3}\right)\Delta f \qquad (2\text{-}8)$$

where R_{B1} is the value of the base resistors of the input stage Q_1, Q_2. Because of the relatively high transconductance of bipolar transistors, the noise voltage of the bipolar input stage is considerably smaller than the noise voltage of a CMOS input stage. If large bias currents of the order of $1mA$ are used, and if large transistors with small base resistance R_{B1} are applied, the noise voltage can be as low as $1nV/\sqrt{Hz}$. The noise corner of the flicker noise component is at such a low frequency in bipolar technology that the flicker noise can be neglected in most applications.

2.3 Rail-to-rail input stages in bipolar and CMOS technology

To make efficient use of the supply voltage at the input of the amplifier for applications as input and output buffers, input stages should be able to handle common-mode input voltages from rail to rail. There are several possibilities to achieve a rail-to-rail common-mode input voltage range. One way is by placing an N-type input stage in parallel with a P-type input stage. Other techniques include using a charge pump to increase the internal supply voltage of a single input pair [2], or using a single input pair consisting of depletion-mode transistors [3]. The last two techniques achieve better common-mode rejection and lower distortion compared to the first technique where the switch-over between the two stages reduces the common-mode rejection.

However, a charge pump requires a considerable die area because of the capacitors required to store the charge and to limit the ripple at the output while depletion-mode transistors are only available in special expensive processes. Therefore, application of a charge pump or a special process can only be justified for specific applications where high accuracy is required in unity-gain configuration. Because of these reasons we limit our discussion to complementary rail-to-rail input stages consisting of an N-type input stage Q_3, Q_4 in parallel with a P-type input stage Q_1, Q_2, as shown in Fig. 2-3 [4]. The outputs of the input stages are added in the summing circuit consisting of Q_5-Q_8 and R_5-R_8. Without a g_m-control circuit Q_{11}-Q_{13}, the transconductance g_m varies a factor of two over the common-mode input voltage range, as shown in Fig. 2-4 by the dotted line. This increases the distortion of the amplifier if applied in a non-inverting configuration and, further, impedes an optimal frequency compensation of the amplifier because the variation of input-stage transconductance results in a variation of the unity-gain frequency of the amplifier. The stability of the amplifier must be guaranteed in the high-transconductance region requiring a factor of two more bias current in the output stage, while the minimum bandwidth specification is determined by the low-transconductance region.

These problems can be solved by inserting a proper g_m-control circuit. In bipolar technology, the transconductance g_m is proportional to the collector current. Therefore, a constant g_m can be obtained by keeping the sum of the tail-currents of the complementary input stages constant as given by

$$I_{Q1,\,Q2} + I_{Q3,\,Q4} = I_1 \tag{2-9}$$

where $I_{Q1,Q2}$ is the bias current of the PNP input pair, $I_{Q3,Q4}$ the bias current of the NPN input pair, and I_1 is the tail current source. This gives the sum of the transconductance of the input pair as

$$g_{m,\,Q1,\,Q2} + g_{m,\,Q3,\,Q4} = g_1 \tag{2-10}$$

where $g_{m,Q1,Q2}$ is the transconductance of the NPN input stage, $g_{m,Q3,Q4}$ is the transconductance of the PNP input stage, and g_1 is the transconductance corresponding to the current I_1. The g_m-control circuit is realized by transistors Q_{11}-Q_{13} as shown in Fig. 2-3 [5]. Depending on the common-mode input voltage the current switch, Q_{11},

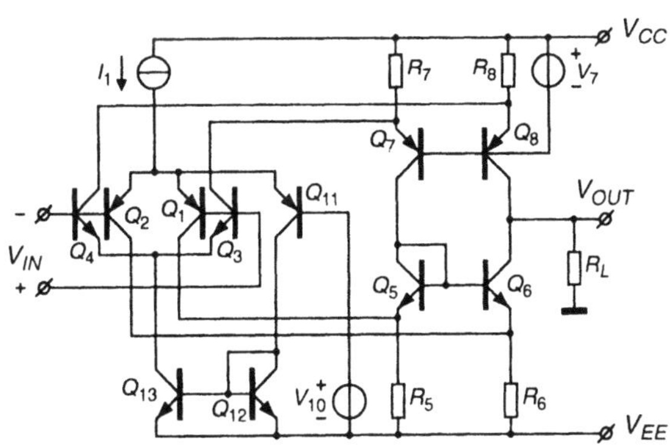

Fig. 2-3 *Rail-to-rail bipolar input stage with g_m-control circuit consisting of a current switch and a current mirror.*

directs the tail-current, I_1, to either one of the input stages. The result is a constant g_m

over the common-mode input range, as shown in Fig. 2-4 by the solid line. For full rail-

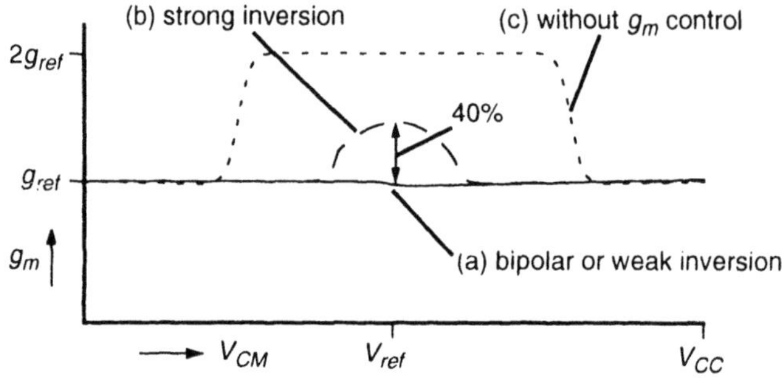

Fig. 2-4 *Transconductance g_m versus common-mode input voltage of a bipolar rail-to-rail input stage with g_m-control circuit (a), of a CMOS rail-to-rail input stage operating in strong inversion using the same g_m-control (b) and of a rail-to-rail input stage without g_m-control (c).*

to-rail operation, the input stages needs a supply voltage that is larger than two base-emitter voltages and two saturation voltages which is about 1.6*V*, depending on the technology. At smaller supply voltages a gap appears in the middle of the common-mode input-voltage range where the input stage does not function.

In CMOS technology, the g_m of a transistor is also proportional to the drain current when it operates in weak-inversion and, therefore, a similar circuit can be used as for bipolar technology. Another possibility is to generate an internal supply voltage for the PMOS input stage so that the switching on of one input stage coincides with the switching off of the other one [6]. However, this technique uses a rather complicated control circuit which consumes a lot of die area.

In strong-inversion the bipolar g_m-control technique leads to a g_m which varies approximately 40% over the common-mode input voltage range, as shown in Fig. 2-4 by the dashed line. The reason for this is that the g_m of an MOS transistor which operates in strong-inversion is proportional to the square-root of the drain-current instead of proportional to it. Proper g_m control based on controlling the input stage bias currents therefore requires a more complicated control circuit that regulates the sum of the square-root of the bias currents in such a way that it is constant [7]. Because the transconductance of CMOS transistors operating in strong inversion is proportional to the gate-source voltage, an alternative to controlling the currents of CMOS rail-to-rail input stages is controlling the voltages[7, 8]. Ideally, a constant voltage source should be connected between the sources of the P-channel and the N-channel pair that equals the sum of a reference gate-source voltage of a PMOS transistor and a reference gate-source voltage of an NMOS transistor. A first approach is shown in Fig. 2-5. In this input stage

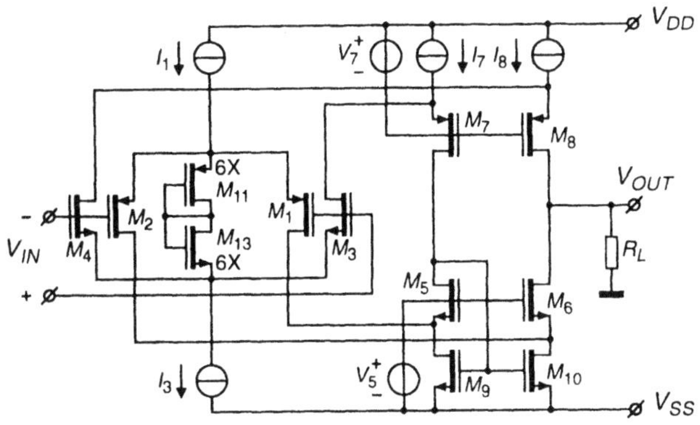

Fig. 2-5 *Rail-to-rail CMOS input stage with constant-voltage g_m control using two diodes.*

the reference voltage is formed by a diode-connected PMOS transistor M_{11} in series with a diode-connected NMOS transistor M_{13}. If the W over L ratio of these transistors is six times the W over L ratio of the input transistors, the current through the diode-connected transistors is $6I_{REF}$ in the intermediate part of the input-voltage range. The remaining current $2I_{REF}$ is used for biasing both input pairs. In the other parts of the common-mode input-voltage range, no current flows through the reference diodes and one of the input stages is biased by $8I_{REF}$. The result is a constant g_m. In the transition regions the current through the diodes changes. Consequently, the voltage across the diodes changes. We can evaluate this effect by examining the translinear loop that is formed by the input stage-transistors and the g_m-control diodes, yielding the following relation

$$\sqrt{I_{1,2}} + \sqrt{I_{3,4}} = 2\sqrt{\frac{I_{11,13}}{3}} \qquad (2\text{-}11)$$

where $I_{1,2}$ is the bias current of the PMOS input stage M_1, M_2; $I_{3,4}$ the bias current of the NMOS input stage M_3, M_4; and $I_{11,13}$ the current flowing through the diodes. At the edges of the region where both input pairs are operating, the bias current of one of the input stages is nearly zero. If, for example, $I_{1,2}$ is zero, the relation between the bias current of the NMOS transistors $I_{3,4}$ and the currents flowing through the diodes can easily be found using equation (2-11). Since the sum of these two currents is equal to

$8I_{REF}$, the bias current of the NMOS transistors can be calculated. Thus, it is found that the transconductance in this operating point is 24% smaller than in the middle of the common-mode range. This can also be concluded from the plot of the transconductance shown in Fig. 2-6. The g_m-control can be further improved by making the current

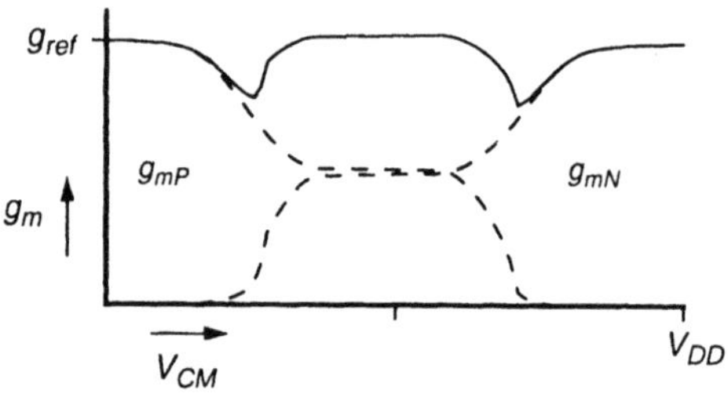

Fig. 2-6 *g_m versus the common-mode voltage for the CMOS rail-to-rail input stage with constant-voltage control.*

through the diodes more constant using a feedback loop [8].

The minimum supply voltage for a full rail-to-rail input range is equal to two gate-source voltages and two saturation voltages which amounts to about 2.5*V*, depending on technology.

A simple feedforward method to control the g_m of the input stage is obtained by applying two current switches as shown in bipolar technology in Fig. 2-7 [9,10]. The advantage of such a feedforward control is that no control loop is needed in which delay would be introduced. Therefore, the bandwidth of a rail-to-rail input stage using this principle can be high. In this circuit the current switches Q_{11}, Q_{13} remove the current that is not needed in such a way that the sum of the tail currents is constant, as can be concluded by evaluating the translinear loop consisting of the PNP input transistors Q_1, Q_2; the PNP current switch Q_{11}; the NPN current switch Q_{13}; and the NPN input transistors Q_3, Q_4. This gives the relation

$$I_{Q1,Q2} I_{Q3,Q4} = (I_1 - I_{Q1,Q2})(I_3 - I_{Q3,Q4}) \tag{2-12}$$

where it is assumed that the emitter area of the current switches is two times the emitter area of the input transistors. If I_3 is equal to I_1, equation (2-9) is obtained. The transconductance as a function of the common-mode input voltage is plotted in Fig. 2-8.

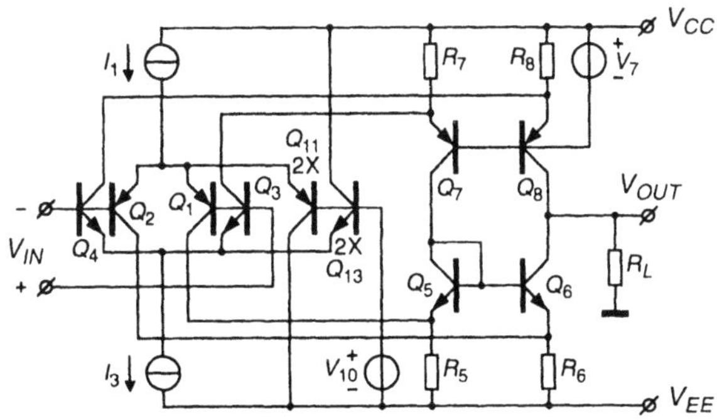

Fig. 2-7 *Rail-to-rail bipolar input stage with g_m control using two current switches.*

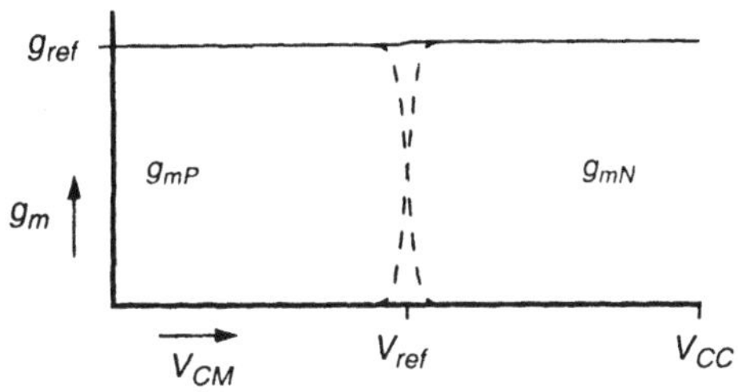

Fig. 2-8 *g_m versus the common-mode voltage for the rail-to-rail input stage with two current switches.*

The same circuit can also be applied in CMOS technology as depicted in Fig. 2-9. Again, a part of the bias current of the input stage is removed by current switches M_{11}, M_{13} in the middle of the common-mode voltage range. For high common-mode input voltages the NMOS input pair operates and the PMOS input pair is off, while the bias current of the PMOS pair completely flows through the switch M_{11}. For low common-mode input voltages, the PMOS input stage functions and all the bias current of the NMOS pair flows through the switch M_{13} so that the NMOS input stage is off. In the middle of the common mode range, when the common-mode voltage is equal to the reference voltage V_{10}, both input stages operate together and, therefore, the g_m of each input stage should be a factor two smaller than when only one input stage functions. In weak inversion the operation of the current switches is the same as in bipolar technology and half the bias current should be removed. In strong inversion the current through the input stage transistors should be a factor 4 smaller, which can be obtained by assuring that 3/4 of the bias current flows through the current switches. This is obtained if the W/L ratio of the

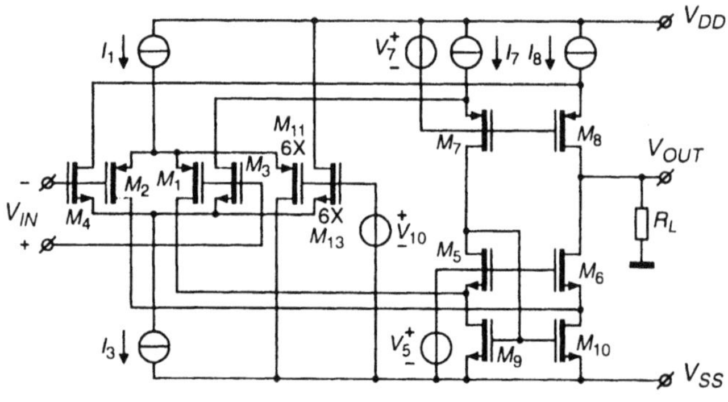

Fig. 2-9 *CMOS rail-to-rail input stage with g_m-control using two current switches.*

current switches is a factor 3 larger than the W/L ratio of both input transistors. The current switches form an MOS translinear loop together with the input-stage transistors which gives the relation

$$\sqrt{I_{1,2}} + \sqrt{I_{3,4}} = \frac{\sqrt{8I_{REF} - I_{1,2}} + \sqrt{8I_{REF} - I_{3,4}}}{\sqrt{3}} \qquad (2\text{-}13)$$

where $I_{1,2}$ is the current flowing through M_1 and M_2, and $I_{3,4}$ is the current flowing through M_3 and M_4. It can be easily verified that the case where $I_{1,2}$ and $I_{3,4}$ are both equal to $2I_{REF}$ satisfies equation (2-13). Another interesting situation is found when one of the currents $I_{1,2}$ or $I_{3,4}$ is zero. In that case, using equation (2-13) it can be calculated that the other current is equal to $6I_{REF}$ instead of $8I_{REF}$ which would be required to obtain constant g_m. This causes a deviation of 13% which can also be concluded from the g_m plot shown in Fig. 2-10. A disadvantage of the switches used in Fig. 2-9 is that the W/L ratio of the current switches is a factor three larger than the sum of the W/L ratios of the two corresponding input-stage transistors. Therefore, a considerable die area is consumed by the g_m-control circuit. If we would apply current switches with a smaller W/L ratio, the gate-source voltage of the current switches in the middle of the common-mode input voltage range would be larger than the desired gate-source voltage. In the next section we discuss how this can be compensated by a level shift inserted between the gates of the two switch transistors.

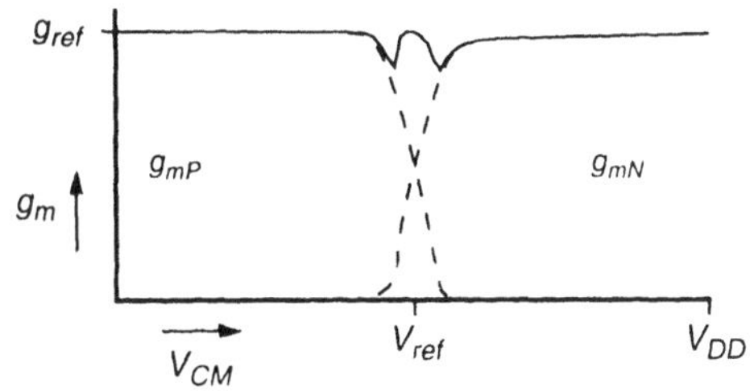

Fig. 2-10 *Transconductance as a function of the common-mode input voltage using the CMOS rail-to-rail input stage with current switches in strong inversion.*

Finally, it should be noted that the offset voltage of rail-to-rail input stages changes when the complementary input stage gradually switches from one input pair to the other. This change of the offset voltage degrades the *CMRR* in the transition regions according to

$$CMRR = \frac{\Delta V_{CM}}{\Delta V_{OS}} \tag{2-14}$$

If, for example, the offset difference between P-type input stage and N-type input stage is 10 *mV*, and the switch-over range is 0.5 *V*, the worst case CMRR is 34 *dB* for small signals at common-mode voltages around the switch-over reference voltage. The CMRR can be improved by increasing the switch-over range or by reducing the offset using calibration techniques.

2.4 Rail-to-rail input stages with constant common-mode output currents

A disadvantage of the rail-to-rail input stages is that the output currents of the input stages vary as a function of the common-mode input voltage. This poses large constraints on the design of the cascoding and summing circuit. If the output current as a function of the common-mode voltage can be made constant, a simple summing circuit can be used yielding a compact implementation of the complete rail-to-rail input stage with summing circuit. The g_m-control using current switches can easily be modified in such a way that the output currents are independent of the common-mode voltage, as shown in Fig. 2-11. Instead of removing the current to the supply rails, the current is passed around the input transistors using two double switches Q_{11}, Q_{12} and Q_{13}, Q_{14}, so that the bias currents in the intermediate stage do not change as a function of the common-mode voltage. The constant bias currents allow a simple and compact implementation of an extra output signal without duplicating the complete rail-to-rail input stage, as indicated by the output at the collector of Q_{16}. Such an extra output is used in multipath compensated amplifiers. A drawback of the circuit is that the current switches add noise which is of the same order as the noise of the input transistors. This can be improved by inserting emitter R_{11}-R_{14}. To compensate for the voltage drop across these resistors, a level shift V_{LS} has to be

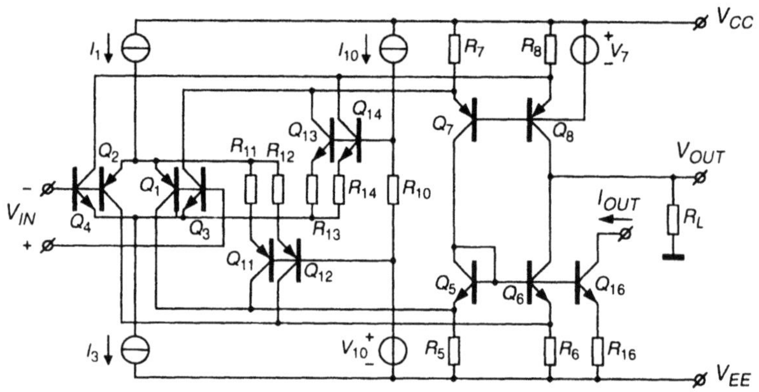

Fig. 2-11 *Rail-to-rail bipolar input stage with g_m control using two current switches and emitter resistors and constant common-mode output currents.*

inserted between the bases of the PNP switches and the NPN switches using resistor R_{10}. Further, the resistors increase cross-over range as shown in Fig. 2-12. This helps to

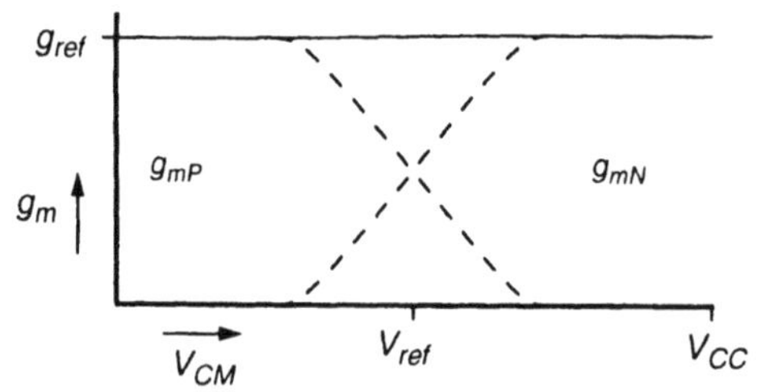

Fig. 2-12 *g_m versus the common-mode voltage for the rail-to-rail input stage with two current switches and emitter resistors.*

reduce the distortion in the cross-over range caused by differences between the offset of the two input pairs. The resistors and level shift can also be used in the bipolar circuit with two switches described in the previous section in order to increase the cross-over range.

The same rail-to-rail input stage with constant common-mode output currents can also be implemented in CMOS technology as shown in Fig. 2-13. By inserting the level shift

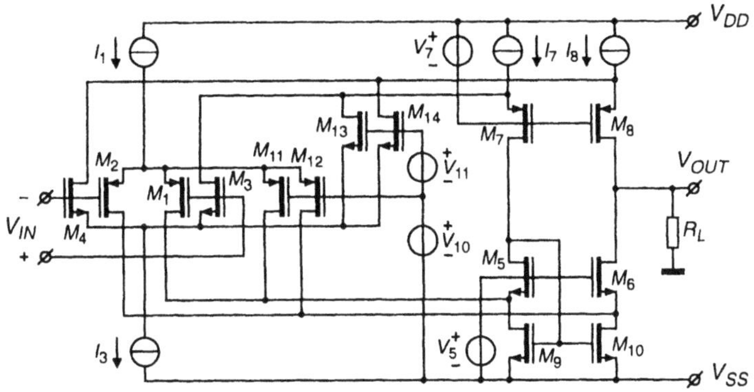

Fig. 2-13 *Rail-to-rail CMOS input stage with g_m control using two current switches.*

V_{11}, the noise contribution of the current switches can also be reduced. Further, the *W/L* ratio of the switch transistors can be made smaller yielding a smaller die area. Again, the same technique may be applied to the CMOS circuit with two switches described in the previous section in order to extend the cross-over range and to reduce the area of the switches.

Another possibility to reduce the noise of the current switches using the rail-to-rail output stage with constant common-mode currents exists in BiCMOS technology, as shown in Fig. 2-14. The input stage consists of a bipolar rail-to-rail input stage Q_1-Q_4 with bipolar switches Q_{11}, Q_{12} that supply their current to the outputs of the input stages using CMOS transistors M_{13}-M_{16}. Thus, the collector currents of the bipolar switches is supplied to both outputs of the corresponding differential pair. Therefore, the bipolar output transistors do not contribute to the noise. The CMOS transistors M_{13}-M_{16} do contribute to the noise, but because their transconductance is small compared to the input-stage transistors their noise contribution is negligible.

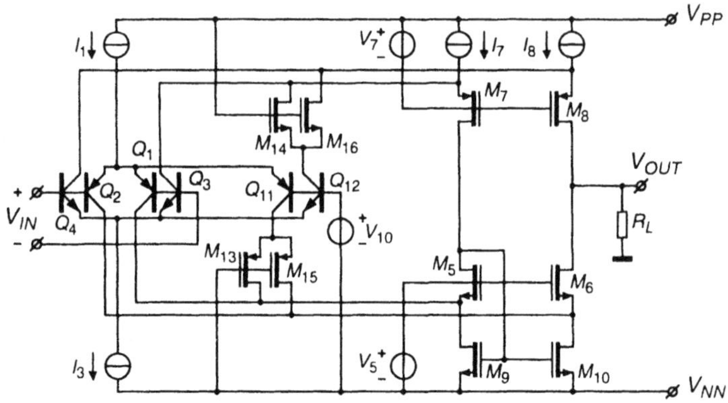

Fig. 2-14 *BiCMOS example of bipolar rail-to-rail input stage with current switches and constant common-mode output currents where the noise contribution of the bipolar switches is reduced using CMOS transistors.*

2.5 Input-stage biasing for constant g_m in bipolar and CMOS technology

The transconductance of the input stage is important because it determines the unity-gain frequency of the operational amplifier, as we later discuss in chapter 4. In most applications the unity-gain frequency should be fixed. Therefore, the biasing of the input stage should be such that the transconductance is constant as a function of temperature. It is well known that this is obtained in bipolar technology by using a PTAT current source to bias the input stage. An example of a popular PTAT current source is depicted in Fig. 2-15 [11]. The circuit consists of a PTAT core Q_1, Q_2, which generates a PTAT voltage across R_1 by making the emitter area of Q_1 n times larger than the emitter area of Q_2. Using current sources Q_3 and Q_4 two equal currents are created that bias Q_1 and Q_2. The value of the bias current is then given by

$$I_{PTAT} = \frac{kT}{q} \frac{\ln(n)}{R_1} \tag{2-15}$$

The value of the current sources is set using a feedback loop consisting of Q_5, current mirror Q_6, Q_3, and the PTAT transistors Q_1, Q_2. The feedback loop requires a compensation capacitor C_{COMP}. Using Q_5, the collector emitter voltage of Q_1 and Q_2 are

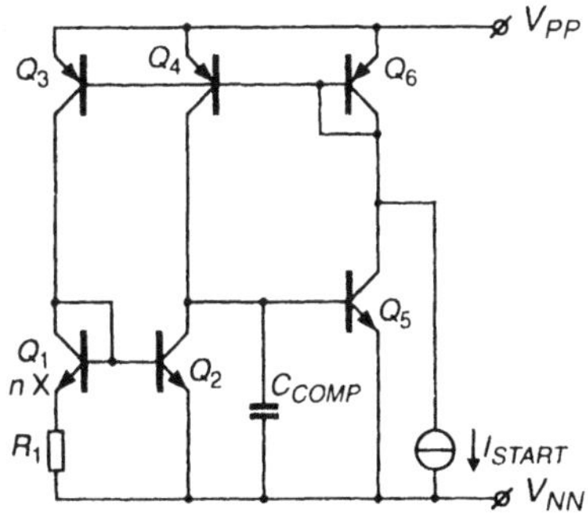

Fig. 2-15 *Low-voltage PTAT current source.*

nearly equal yielding a good power-supply rejection at low frequencies. At higher frequencies the power-supply rejection is reduced by the imbalance caused by C_{COMP}.

A different way of biasing the PTAT transistors is demonstrated in the circuit shown in Fig. 2-16 [12]. The circuit employs a folded NPN current mirror Q_3, Q_4 to force equal

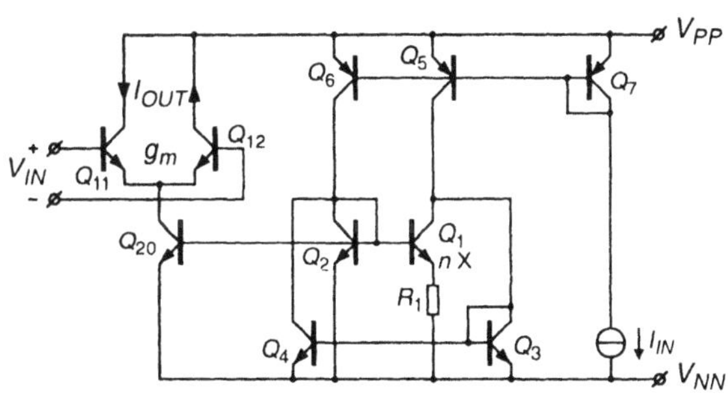

Fig. 2-16 *Folded PTAT current source creates bias current of bipolar input stage for obtaining constant* g_m.

currents in PTAT transistors Q_1 and Q_2. The mirror and the PTAT transistors are biased by two equal current sources Q_5 and Q_6. The value of the current generated by Q_5 and Q_6 is determined by an input current that should be larger than the PTAT current to be generated. In that case, a PTAT current as expressed by equation (2-15) flows through Q_1 and Q_2, while the remaining part of the input current flows through the current mirror Q_3, Q_4. An advantage of the circuit is that no feedback is used so that a compensation capacitor is not needed. Further, high power supply rejection is achieved due to the inherent symmetry. A drawback of the circuit is that an input current I_{IN} is needed that is relatively constant as a function of the supply range, and that more noise is generated by placing the current mirror in parallel with the PTAT transistors. If the PTAT current is used to bias an input stage Q_{11}, Q_{12} as shown in Fig. 2-16, the transconductance of the input stage is given by

$$g_m = \frac{q}{kT} I_C \propto \frac{\ln(n)}{R_1} \qquad (2\text{-}16)$$

where n is the emitter ratio between transistors Q_1 and Q_2. The transconductance is completely determined by R_1 and not directly dependent on temperature. Thus, the

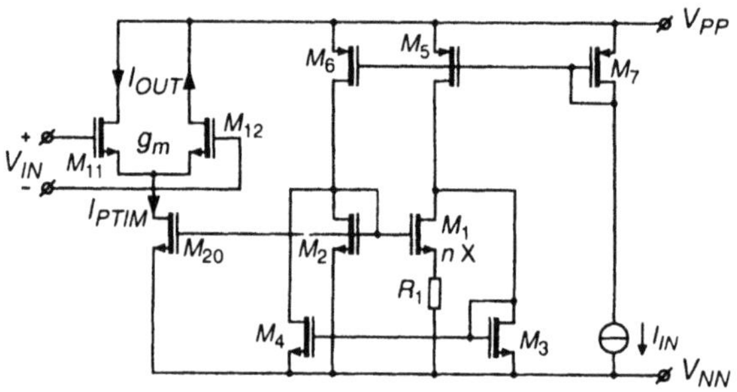

Fig.2-17*CMOS version of folded PTAT current source generates bias current of NMOS input stage for obtaining constant* g_m.

remaining temperature dependency is determined by the type of resistor used.

The same biasing technique can also be applied in CMOS technology as depicted in Fig. 2-17. If the transistors are operating in weak inversion, similar equations are found as in bipolar technology. In strong inversion the principle still functions properly, but different equations are found. The bias current generated in the current source is Proportional To the Inverse of the Mobility (PTIM) of the charge carriers and given by

$$I_{PTIM} = \frac{1}{R_1^2 \mu_n C_{ox}} \frac{2L_1}{W_1} \frac{(\sqrt{n}-1)^2}{n} \qquad (2\text{-}17)$$

If we use I_{PTIM} as bias current of NMOS transistors, the transconductance is given by

$$g_m = \sqrt{\mu_n C_{ox} \frac{W_1}{L_1} I_D} \propto \left(1 - \frac{\sqrt{n}}{n}\right) \frac{1}{R_1} \tag{2-18}$$

where n is the ratio between the W/L ratio of M_1 and the W/L ratio of M_2. Again, the transconductance is completely determined by resistor R_1. Of course, the transistors of the current source should operate in the same region by assuring that the current densities of the input transistors are approximately equal to the current densities of the current source. Also, they should be of the same type to avoid influence of the ratio μ_n/μ_p.

Finally, we need to implement the input current of the folded PTAT current source. Any circuit that generates a current which is reasonably constant over the supply-voltage range can be applied. Examples included conventional PTAT current sources. This can be a central current source that supplies many local folded sources on a VLSI chip. Another possibility is to apply a peaking current source as demonstrated in the BiCMOS example shown in Fig. 2-18. The peaking current source consists of a current mirror Q_9, Q_8 of which the bias current is set by resistor R_{10}. Thus, the current I_{IN} generated by the peaking-current source increases with the supply voltage. At a certain moment, however, the output current decreases again due to the voltage drop across R_9, yielding a flat area in the middle of the supply-voltage range. This behavior is depicted in Fig. 2-19 where the current I_{IN} generated by the peaking-current source is plotted as a function of the supply voltage together with the output current I_{PTAT} of the folded PTAT source. The current mirrors of the folded PTAT current source are implemented in CMOS technology to avoid influence of the base currents.

2.6 Conclusions

In this chapter we discussed how the common-mode input-voltage range of input stages in CMOS and bipolar technology can be increased using compact rail-to-rail input stages. A g_m-control circuit is needed that regulates the transconductance of the input stage constant because otherwise the bandwidth and gain of the amplifier would vary as a function of the common-mode voltage. This would increase the distortion and hinder an optimal frequency compensation. Another disadvantage of rail-to-rail input stages is that the common-mode rejection decreases because of the offset that changes when the input switches from N-type input stage to P-type input stage. This can be slightly improved by increasing the switch-over region. Finally, we have seen how a PTAT current source can be used to bias bipolar input stages in order to obtain a stabilized transconductance over

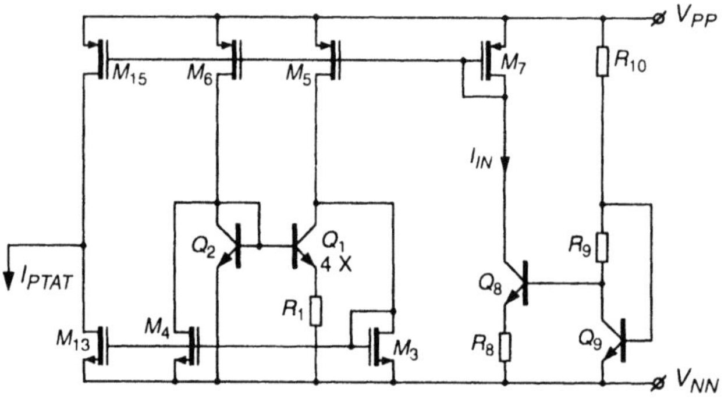

Fig. 2-18 *Complete BiCMOS implementation of folded PTAT current source.*

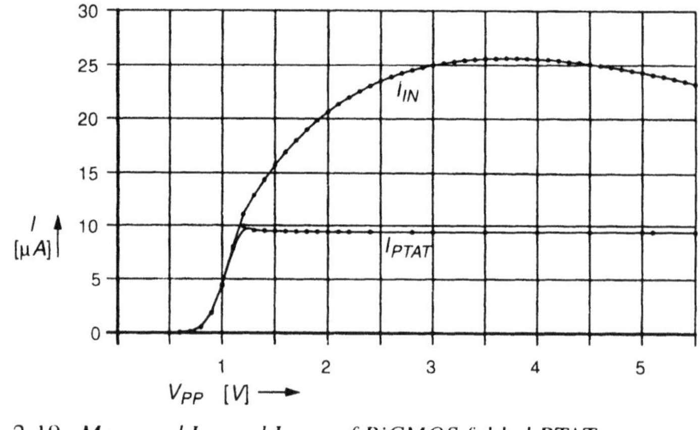

Fig. 2-19 *Measured I_{IN} and I_{PTAT} of BiCMOS folded PTAT current source.*

the temperature range. In CMOS the same technique can be used yielding a stabilized transconductance in weak inversion as well as in strong inversion.

2.7 References

[1] M.J.M. Pelgrom, A.C.J. Duinmaijer, A.P.G. Welbers, "Matching Properties of MOS Transistors", *IEEE J. Solid-State Circuits*, vol. 21, no. 6, December 1986, pp. 1433-1439.

[2] T.A.F Duisters, E.C. Dijkmans, "A -90-dB THD Rail-to-Rail Input Opamp Using a New Local CHarge Pump in CMOS", *IEEE J. Solid-State Circuits*, vol. 33, no. 7, July 1998, pp. 947-955.

[3] R. Griffith, R.L. Vyne, R.N. Dotson, T. Petty, "A 1-V BiCMOS Rail-to-Rail Amplifier with n-Channel Depletion Mode Input Stage", *IEEE J. Solid-State Circuits*, vol. SC-32, no. 12, pp. 2012-2022, Dec. 1997.

[4] J.H. Huijsing, R.J. v.d. Plassche, "Differential Amplifier with Rail-to-Rail Input Capability and Controlled Transconductance", U.S. Patent, No. 4, 555, 673, Nov. 26, 1985

[5] J. Fonderie, J.H. Huijsing, *Design of Low-voltage Bipolar Operational Amplifiers*. Boston: Kluwer Academic Publishers, 1993.

[6] G.Ferri, W.M.C. Sansen, "A Rail-to-Rail Constant-gm Low-Voltage CMOS Operational Transconductance Amplifier", *IEEE J. Solid-State Circuits*, vol. 32, no. 10, October 1997, pp. 1563-1567.

[7] R. Hogervorst, J.H. Huijsing, *Design of Low-voltage Low-Power Operational Amplifier Cells*. Dordrecht, The Netherlands: Kluwer Academic Publishers, 1996.

[8] R. Hogervorst, J.P. Tero, J.H. Huijsing, "Compact CMOS constant-g_m rail-to-rail input stage with g_m-control by an electronic zener diode", *IEEE J. Solid-State Circuits*, vol. 31, no. 7, July 1996, pp. 1035-1040.

[9] J.H. Huijsing, R. Hogervorst, "Rail-to-rail input stage with constant-g_m and constant common-mode output currents", US pat. 5734297, March 31, 1998.

[10] K.J. de Langen, R. Hogervorst, J.H. Huijsing, "Translinear circuits in Low-Voltage Operational Amplifiers", in *Advances in Analog Circuit Design: MOST RF Circuits, Sigma-Delta Converters and Translinear Circuits*, W.M.C. Sansen, J.H. Huijsing, R.J. van de Plassche, Eds. Dordrecht, The Netherlands: Kluwer Academic Publishers, 1996, pp. 357-38.

[11] H.C. Nauta and E.H. Nordholt, "New Class of High-Performance PTAT Current Sources", Electron. Lett., vol. 21, pp. 384-386, April 1985.

[12] K.J. de Langen, J.H. Huijsing, "Compact Low-Voltage PTAT-current source and Bandgap-Reference Circuits", *Proc. European Solid-State Circuits Conference*, The Hague, The Netherlands, September 22- 24, 1998, pp. 108-111.

Output Stages 3

3.1 Introduction

The main function of the output stage is to deliver the output signal to the load. Since high signal levels are likely to occur, the output stage should be able to deal with high voltages and high currents. For a good power efficiency the maximum peak-to-peak output voltage should be nearly equal to the supply voltage and the stage should be biased by a very small quiescent current while being capable of driving much larger output currents. Also, the output stage has to cope with the load connected to the output terminal. The resistive part of the load determines the maximum output current that is needed. Very low resistive loads in the order of 50 ohms require a maximum output current of at least 10 mA and pose large demands upon the construction of the output stage. For example, large output transistors must be used which result in large parasitic capacitances. Consequently, the attainable bandwidth is reduced by the larger capacitances. The capacitive load, on the other hand, determines the minimum current, or quiescent current, that is needed to obtain sufficient phase margin and stability. Finally, distortion is an important overall parameter that is mainly determined by the output stage, because it handles the highest signal levels.

To satisfy all the above requirements a compromise must be reached between power efficiency on one hand and distortion and stability on the other. The implementation of such a compromise is obtained by using class-AB biasing.

We start by examining the basic configurations that can be used to construct our output stages in section section 3.2. The distortion of output stage is discussed in section 3.3 Thereafter, class-AB biased rail-to-rail output stages using feedforward control are presented in section 3.4 and feedback biased class-AB output stages are discussed in section 3.5. All-NPN output stages are presented in section 3.6 and, BiCMOS output-stage configurations options are discussed in section 3.7. Circuits to protect the output stages are discussed in section 3.8 and, finally, conclusions are drawn in section 3.9.

3.2 Basic output-stage configurations

All the basic single-stage transistor stages, the voltage follower, the current follower, and the inverting amplifier, can in principle be used as output stage. Most important are the voltage follower configuration and the inverting amplifier configuration as shown in bipolar technology in Fig.3-1. Current followers or cascodes may also be used as output

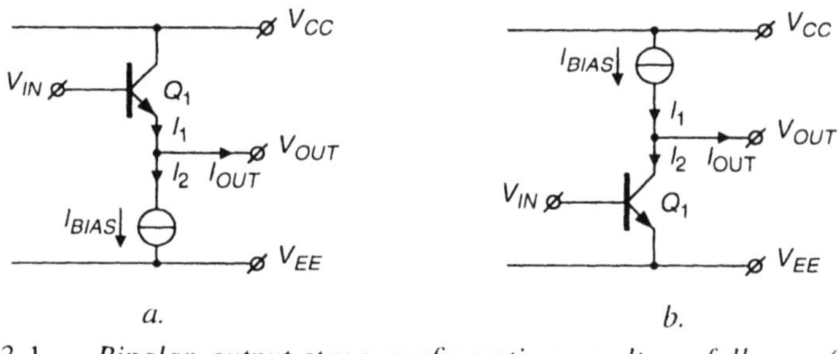

Fig. 3-1 *Bipolar output-stage configurations: voltage follower (emitter follower or CC stage) (a) and inverting amplifier (CE stage) (b).*

transistors. However, a current follower does not provide current gain and therefore the handling of large output currents must be solved in a preceding inverting amplifier stage which can be regarded as the output stage in that case. Further, the output-voltage range of a cascode output transistor is reduced by the voltage required across the cascode which can often not be allowed at low supply voltages. For these reasons current followers are not discussed in this chapter.

The disadvantage of the single-transistor output stages is that the output current in one direction is limited to the value of the current source. The current that can be pulled out of the load by the voltage follower is limited to I_{BIAS} and the current that can be pushed into the load by the inverter is also limited to I_{BIAS}. Thus, using single transistors, a high power efficiency cannot be obtained. By using two transistors however, high push and pull currents can be combined yielding a push-pull output stage. By combining complementary versions of the voltage follower or the inverting amplifier, or by combining the voltage follower and the inverting amplifier, four push-pull configurations can be constructed of which the three most important are shown in Fig.3-2.

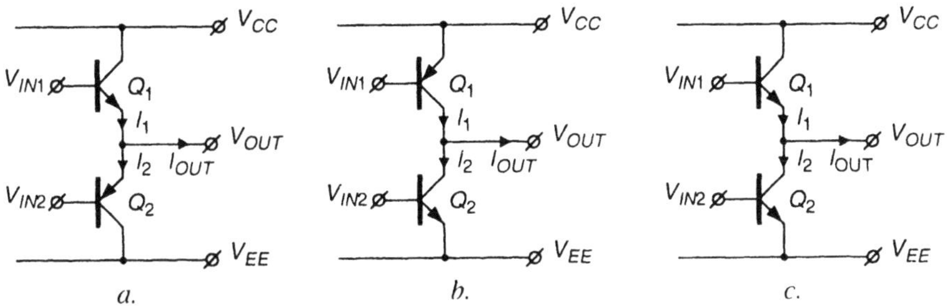

Fig. 3-2 *Bipolar push-pull output-stage configurations: complementary voltage follower (a), complementary inverting amplifier (b) and, All-NPN output stage (c).*

The first output stage configuration is the complementary voltage follower as depicted in Fig.3-2a, consisting of two output transistors, Q_1 and Q_2, connected in a common-collector configuration. This circuit forms the basis for the well-known complementary class-AB output stage which has been applied in many famous opamp designs such as the µA741. Today, the small output voltage range which is equal to the supply voltage minus the two base-emitter voltages of the output transistors renders this stage useless in low-voltage designs. However, because it can be simply biased, it is a good starting point for constructing output stages based upon the other configurations. Fig.3-2b shows the complementary inverting amplifier configuration. This output stage has the highest possible output voltage range which is equal to nearly the whole supply voltage and is therefore known as rail-to-rail output stage. But, the use of the PNP output transistor poses some limitations, especially when only lateral PNPs are available. First of all, the low transit frequency f_T of lateral PNPs limits the bandwidth to about 10 MHz. Second, due to the low current gain β_F, the output current is limited to mediocre values unless the preceding stage is also designed to deliver large currents. For these reasons, the all-NPN output stage which is shown in Fig.3-2c is still an attractive alternative. Thanks to using only NPN transistors, the bandwidth and the maximum output current can be high. The output voltage range is equal to the supply voltage minus one base-emitter voltage. At a supply voltage of 1.8 *V*, the output voltage range is about half the output voltage range of the inverter but this can be easily compensated by the much higher bandwidth or the much higher output current capability. Theoretically, a fourth possible configuration would be a PNP version of the all-NPN output stage. However, because in most processes the performance of the NPN is much better than the performance of the PNP transistor, an all-PNP output stage is of little use. Deriving the MOS push-pull output-stage configurations from the bipolar circuits is straightforward. Due to the quadratic relation between gate-source voltage and drain current, the gate-source voltage of CMOS transistors is much higher at high output currents than the base-emitter voltage of bipolar transistors. Therefore, we cannot allow the reduction of the output voltage range by a gate-source voltage, furthermore only the complementary CMOS inverting amplifier is important in low-voltage design.

Table 3-1 *Comparison of low-voltage output stage configurations in bipolar and MOS technology.*

	Bipolar rail-to-rail output stage	CMOS rail-to-rail output stage	All-NPN output stage
Circuit	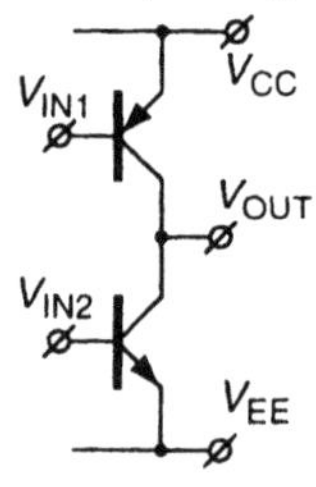	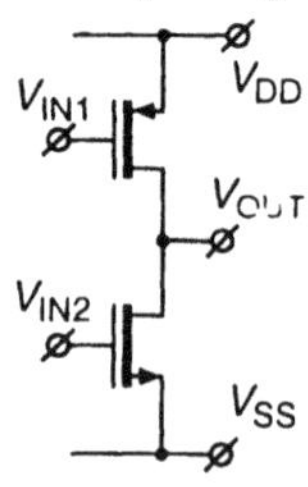	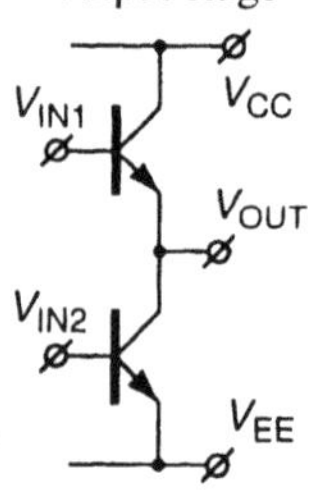
Min. supply voltage	1.0 V	0.9 ··· 1.2 V	2 V
Output voltage range	$V_{CC} \cdots V_{EE}$	$V_{DD} \cdots V_{SS}$	$V_{CC} - V_{BE} \cdots V_{EE}$
Max. output current	20 mA	3 ··· 10 mA	10-1000 mA
Current gain	50/100	∞	100
Transconductance	$\propto I_C$	$\propto \sqrt{I_D}$	$\propto I_C$
Bandwidth	1-100 MHz	1-100 MHz	10-1000 MHz

To summarize the possible configurations and compare their performance, the circuits suited for low-voltage operation are shown together in Table3-1 and an overview of the most important output-stage parameters is given. The values listed in Table3-1 should be considered as an indication. Exact values depend on the process, the bias current, the dimensioning of the output stage, and the context of the circuit. For amplifiers operating at very low supply voltages, the inverting amplifier stages should be applied. These circuits display an output voltage range that is nearly equal to the supply voltage and can operate on supply voltages down to about $1V$. The minimum supply voltage of the CMOS circuit depends on the threshold voltages of the process and on the maximum output current due to the quadratic characteristic. The performance of the bipolar rail-to-rail output stage suffers from the inferior behavior of the PNP transistor which reduces the bandwidth to average values. High output currents can only be allowed at high costs. A very large PNP transistor is required which would consumie vast amounts of die area and yield large parasitic capacitance that further reduces the bandwidth. The MOS stage mainly suffers from the quadratic MOS characteristic yielding low maximum output currents at low supply voltages and lower transconductance than the bipolar circuits. Since the transconductance of the output transistors together with the load capacitor determine the output pole, only a moderate bandwidth can be obtained. An advantage over the bipolar stages, however, is that the current gain of MOS transistors is very high so that the bias current of a preceding stage does not need to be considered when designing the maximum output current. In bipolar circuits special care should be taken that there is sufficient current available in the preceding stage in order to deliver the large base current of the output stage, especially when driving a PNP output transistor. Due to the superior quality of NPN transistors, the all-NPN output stage performs very well. However, because the output voltage range is reduced by the base-emitter voltage of the upper NPN, a higher supply voltage is required in order to obtain a sufficiently large output voltage range. In spite of that it is still extremely efficient.

The next problem is how to bias these push-pull stages. This subject is discussed in the subsequent sections.

3.3 Static distortion considerations

Before constructing circuits for biasing of the push-pull stages we will first take a look at the requirements of the biasing of the output stage. The main parameters of the output stage are the power efficiency, the ability to drive capacitive loads without oscillations, and the distortion. Somehow we have to find an acceptable compromise between these parameters. Therefore, we first investigate the sources of distortion in the output stage. Thereafter suitable biasing characteristics can be generated and circuits to implement the desired behavior can be devised.

The simplified circuit diagram of a two-stage amplifier which is shown in Fig.3-3a in bipolar technology and in Fig.3-3b in MOS technology is used to explain the basics of the output-stage related distortion. Although a two-stage circuit is used as an example, the results found in this section are not limited to two-stage amplifiers and can easily be applied to multi-stage amplifiers. The circuit consists of a voltage-controlled current

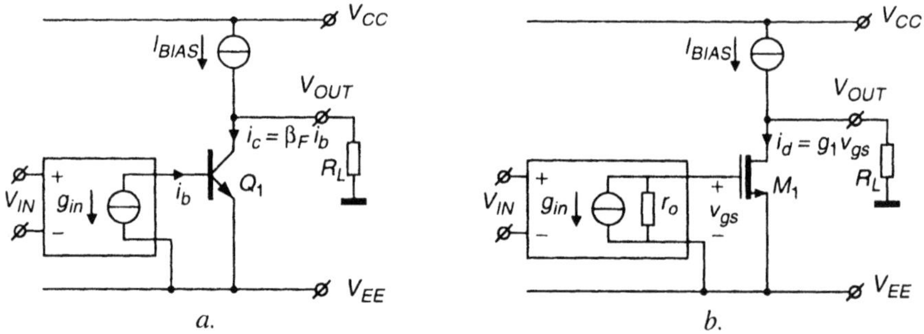

Fig. 3-3 *Simplified diagram of output stage incorporated in a two-stage amplifier in bipolar technology (a) and in CMOS technology (b).*

source with transconductance g_{in} which models the input stage, and an output transistor Q_1 in the bipolar circuit or M_1 in the MOS example. In the bipolar case the DC gain A_0 from input voltage to output voltage is given by

$$A_0 = \frac{V_{OUT}}{V_{IN}} = g_{in}\beta_F R_L \qquad (3\text{-}1)$$

where β_F is the current gain of output transistor Q_1, and R_L is the load resistor. The DC gain of the MOS example is given by

$$A_0 = \frac{V_{OUT}}{V_{IN}} = g_{in} r_{o,\,in} g_1 R_L \qquad (3\text{-}2)$$

where $r_{o,in}$ is the output impedance of the input stage, and g_1 the transconductance of M_1. As we see, the bipolar transistor is current driven for low frequencies and β_F is the parameter that expresses the influence of the output stage. This is even true in cases where a voltage follower is used as output stage because the output impedance of the preceding stage is usually very high. In MOS the output transistor is driven by the voltage generated across the output impedance of the input stage, and the influence of the output stage is dominated by g_1. Therefore, in order to examine the distortion of the output stage, the variations of β_F and g_1 resulting from changes in the output current have to be studied. The current gain β_F of bipolar transistors as a function of collector current I_C is plotted in Fig.3-4a and, the transconductance g_m of an MOS transistor in strong inversion as a function of the drain current I_D is plotted in Fig.3-4b. If the output stage consists of a single transistor and a moderate current I_{BIAS} is used as indicated in the figure, the variation of β_F and g_m is small because the characteristics are relatively flat around I_{BIAS}.

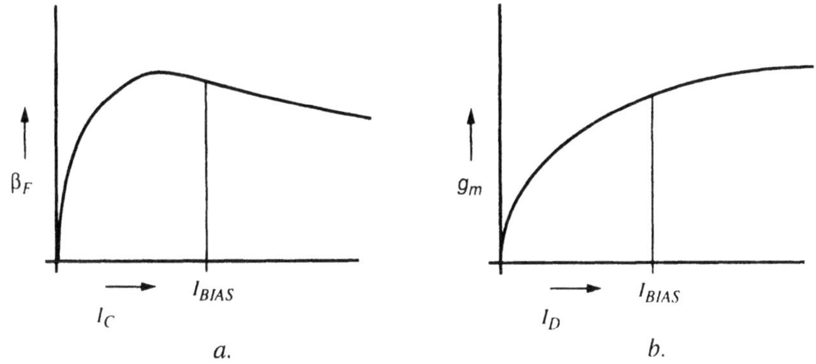

Fig. 3-4 *Distortion in single transistor stages in bipolar technology (β_F-induced non linearity) (a) and using MOS technology (g_m-induced non linearity) (b).*

Additionally, if the current to be delivered to the load is small compared to the bias current, so that only a small region of the characteristics around I_{BIAS} is used, the variation of the parameters can be in the order of 1%. Combined with an amplifier open-loop gain of 80 dB, signal-to-noise ratios of 120 dB or more can easily be obtained. Such a method of biasing is known as class-A biasing. Although showing superior distortion performance, the power efficiency of a class-A stage is very poor and therefore class-AB biasing cannot be used in output stages of general purpose amplifiers and certainly not in modern low-voltage low-power electronics.

Since we are interested in good power efficiency, push-pull stages that can deliver high output currents but only consume very little standby or quiescent current should be examined. A simplified circuit diagram of a rail-to-rail push-pull stage is shown in Fig. 3-5. The voltage sources represent a circuit for controlling the bias current of the output stage. In an extreme case, a stage without any bias current could be considered yielding perfect power efficiency. The current-gain curve of such a stage as a function of output current I_{OUT} is plotted in Fig.3-6a. First of all, we see that the plot is not symmetrical because in most push-pull stages two transistors of complementary type are used. Please note that the all-NPN push-pull stage which, at first glance, appears to be completely asymmetric (because it consists of a voltage follower connected transistor and an

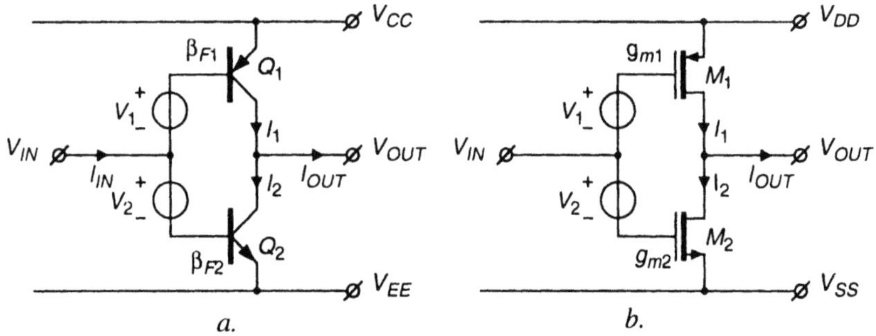

Fig. 3-5 *Simplified diagram of rail-to-rail push-pull output stage in bipolar technology (a) and CMOS technology (b) with voltage sources representing the bias control.*

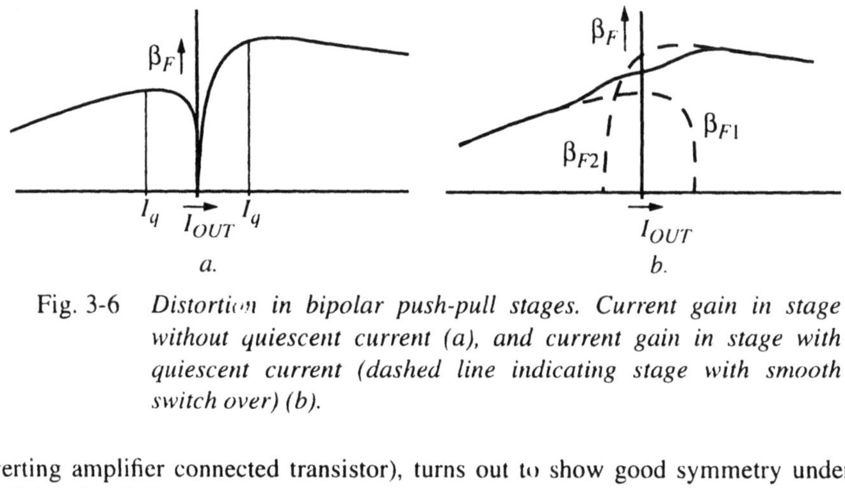

Fig. 3-6 *Distortion in bipolar push-pull stages. Current gain in stage without quiescent current (a), and current gain in stage with quiescent current (dashed line indicating stage with smooth switch over) (b).*

inverting amplifier connected transistor), turns out to show good symmetry under the condition that it is driven by a current source with sufficient high output impedance. With a complementary output stage at moderate values of the collector current however, the current gain β_F is in the order of 100 for NPN transistors and 50 for PNPs. Secondly, the gain changes very rapidly at values of the output current around zero. This is extremely unfortunate, since small signals around the common-mode output voltage are heavily influenced. A push-pull stage without bias current is known as a class-B stage. It exhibits good power efficiency but, the distortion is high, especially for small signal voltages.

The distortion characteristic of MOS stages is comparable to the bipolar case but the shape of the transconductance follows a square-root function, as shown in Fig.3-4, and it displays better symmetry since the difference between the g_m of N-type and P-type transistors is smaller than the difference between the β_F of NPN and PNP transistors. The symmetry depends on the matching of the mobility of electrons μ_n and holes μ_p. The mobility of electrons is about a factor three larger than the mobility of holes. We can equalize the g_m of the PMOS and the NMOS transistor by increasing the W/L ratio of the PMOS by a factor of three. The resulting matching is not very good but can be estimated at approximately 20% which is better than the factor two found in the bipolar output stage. Nevertheless, a rapid variation is also found in the g_m of MOS stages at low current levels which is unacceptable.

The low current region of β_F and g_m can be avoided by enforcing a quiescent current I_Q as indicated in the β_F curve shown in Fig.3-6a by the thin lines. The part of the curve between the thin lines is not used and the plot shown in Fig.3-6b results. Still, the gain changes when the output current crosses zero, but it is already much better than without quiescent current. Further improvement is obtained by gradually switching over from one output transistor to the other output transistor as suggested by the solid line in Fig.3-6b. This way of biasing is known as class-AB biasing and yields a compromise with good power efficiency and acceptable distortion.

Concluding, distortion at low frequencies, or static distortion can be minimized by equalizing the DC gain of the two output transistors and by gradually switching over from one output transistors to the other. In bipolar technology the gain at low frequencies is determined by β_F which cannot be tuned. The only thing left to do is to choose a quiescent current at which the β_F curve is relatively flat. In MOS technology the gain is determined by g_m, and a proper W/L ratio can be selected so that the P-type transistor has approximately the same g_m as the N-type transistor.

As we have seen in this section, a proper class-AB characteristic is an important tool in order to combine low distortion with good power efficiency. Therefore, in the next section we will take a closer look at the theoretical background of class-AB characteristics.

3.4 Feedforward class-AB biasing

The requirements for the shape of the class-AB characteristic were derived in the previous section from a discussion of distortion. It was demonstrated that a smooth switch over should be implemented in order to minimize distortion. There are basically two methods to implement the class-AB biasing of output stages. The first technique is called feedforward control and is discussed in this section. The second method uses a feedback loop to control the biasing of the output stage and is therefore called feedback control. This technique is discussed in the subsequent section.

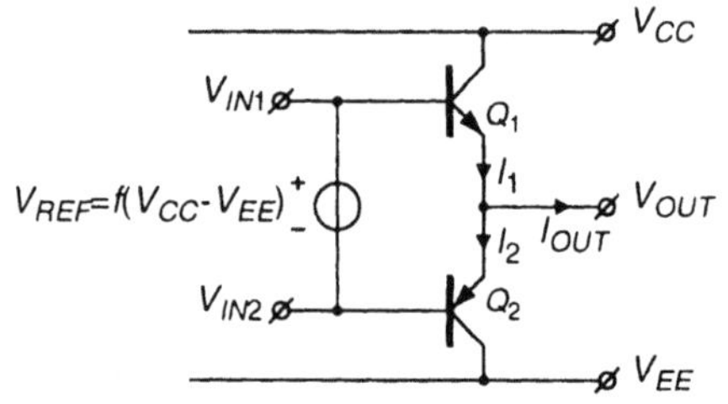

Fig. 3-7 *Feedforward class-AB control principle applied to a complementary output stage.*

The feedforward biasing principle is shown in Fig.3-7 where it is applied to a complementary bipolar output stage. The technique consists of a reference voltage V_{REF} that is directly used to drive the sum of the base-emitter voltages of the two output stages. To obtain a supply-voltage independent quiescent current, V_{REF} must track the supply voltage. Because the base-emitter voltages of the two output transistors of the complementary output stage are connected together, the technique is especially suited to application in the complementary output stage. The output stage can be driven by two input voltages V_{IN1} and V_{IN2}. However, due to the strong coupling between the bases of the output transistors which is the result of voltage source V_{REF}, the circuit can be driven by only one input signal. The technique is not directly applicable to rail-to-rail output stages because the base-emitter voltages of a rail-to-rail output stage are not connected together. However, modification of the basic feedforward principle allows application in rail-to-rail output stages and is discussed at the end of this section. The implementation of the class-AB output stage can be completed by selecting a suitable translinear loop [1] and by modifying the translinear loop in such a way that it comprises the base-emitter voltages of the output transistors and the reference voltage. Inserting a translinear loop is straightforward when the a simple translinear circuit is applied, yielding the well-known bipolar complementary output stage as shown in Fig.3-8 on the left-hand side. Due to the strong coupling by the diodes, the output stage can be driven at one of the bases or, as shown in the figure, at the connection between the two diodes, yielding an input voltage which has the same level as the output voltage.

The biasing of the output transistors is controlled by the translinear loop consisting of the base-emitter voltages of the output transistors Q_1 and Q_2, and the diodes D_3 and D_4. These diodes keep the sum of the base-emitter voltages of Q_1 and Q_2 constant, which

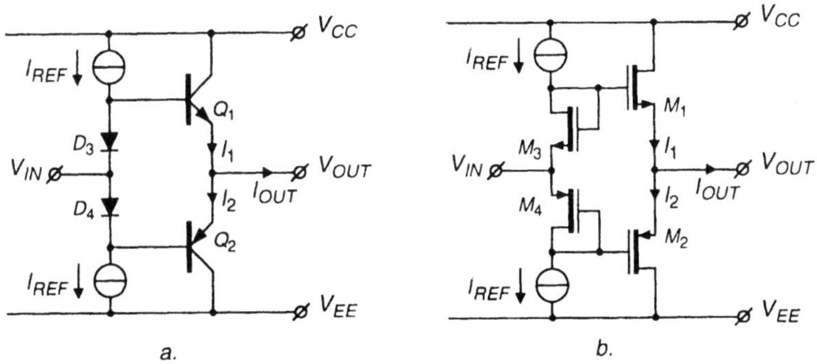

Fig. 3-8 *Conventional feedforward-biased complementary class-AB output stage in bipolar technology (a) and CMOS technology (b).*

results in a constant product of their collector currents, as described by equation (3-19) in the previous section, while the quiescent current I_Q is given by

$$I_Q = \sqrt{\frac{A_1 A_2}{A_3 A_4}} I_{REF} \tag{3-3}$$

in which I_{REF} is the reference current that flows through the diodes, and A_1- A_4 are the emitter area of the corresponding transistors Q_1, Q_2 and diodes D_3, D_4. The class-AB characteristic of the conventional output stage is shown in Fig.3-30. A CMOS version of the complementary output stage is shown in Fig.3-8 on the right-hand side. The class-AB characteristic of the CMOS circuit is described by equation (3-20) and plotted in Fig.3-31. The quiescent current is given by

$$I_Q = \left(\frac{W}{L}\right)_3 \left(\frac{L}{W}\right)_1 I_{REF} \tag{3-4}$$

in which $(W/L)_1$ is the width over length ratio of M_1, and $(W/L)_3$ is the width over length ratio of M_3. Further, it is assumed that the ratio between the width over length ratios of M_2 and M_4 is equal to the ratio between the width over length ratios of M_1 and M_3, and their transconductance values obey

$$K\frac{W}{L} = \frac{1}{2}\mu_n C_{ox}\left(\frac{W}{L}\right)_n = \frac{1}{2}\mu_p C_{ox}\left(\frac{W}{L}\right)_p \tag{3-5}$$

where μ_n is the mobility of electrons, μ_p the mobility of holes, $(W/L)_n$ the width over length ratio of M_1 and M_3, and $(W/L)_p$ the width over length ratio of M_2 and M_4, respectively. In most output stages the latter condition is met, because it reduces the distortion of the amplifier as discussed in section 3.3.

The conventional output stage is not useful in modern low-voltage electronics due to the limitation of the output-voltage swing. The output-voltage swing is limited to the supply voltage minus two base emitter voltages and two saturation voltages in bipolar technology, while in CMOS technology the output-voltage swing is limited to the supply voltage minus two gate-source voltages and two saturation voltages. Therefore, rail-to-rail output stages should be used consisting of two complementary output transistors connected in a common-emitter or common-source configuration, yielding a near rail-to-rail output swing. As discussed above, the separate feedforward biasing principle is useful for application in these circuits. However, implementing the separate feedforward biasing circuit is more complicated than the implementation of the conventional output stage. An easier way to derive a rail-to-rail class-AB biased output stage is by starting with the conventional output stage shown in Fig.3-8. From the conventional circuit with common-collector connected transistors, a rail-to-rail output stage with common-emitter connected transistors can be obtained by interchanging the lower part of Fig.3-8 consisting of Q_2 and D_4 and the upper part, consisting of Q_1 and D_3. The result is shown in Fig.3-9. The translinear loop that controls the current in the output transistors consists

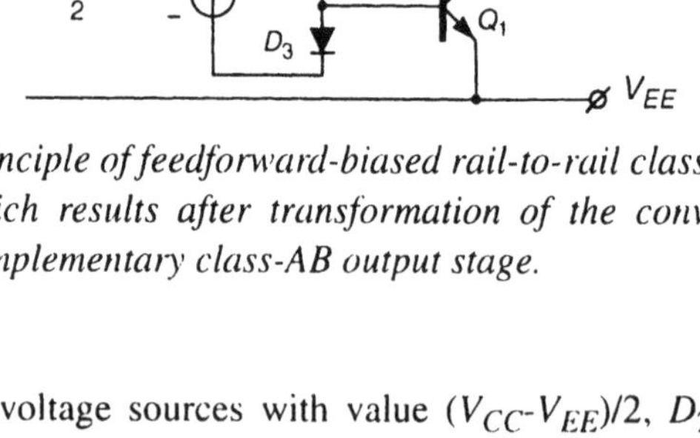

Fig. 3-9 *Principle of feedforward-biased rail-to-rail class-AB output stage which results after transformation of the conventional bipolar complementary class-AB output stage.*

of Q_1, D_3, the two voltage sources with value $(V_{CC}$-$V_{EE})/2$, D_2, Q_2, and the supply voltage V_{CC}-V_{EE}. If the sum of the two voltage sources is equal to the supply voltage, the sum of the base-emitter voltages of the output transistors is equal to the sum of the diode voltages, D_3 and D_4, and the conventional class-AB behavior is exactly implemented.

However, the two floating voltage sources are difficult to implement. The principle can be simplified yielding the circuit shown in Fig. 3-10. In this circuit the reference voltage

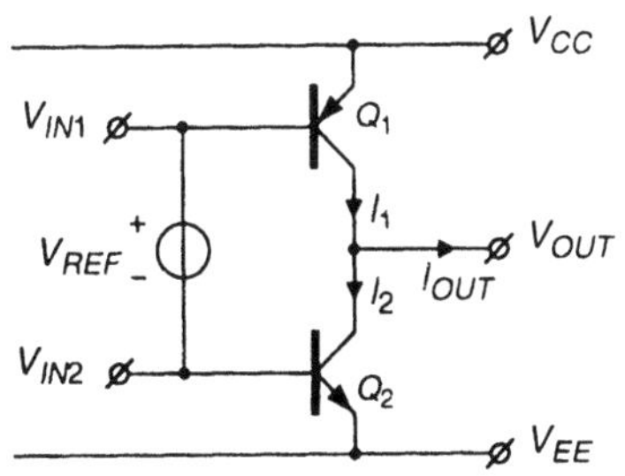

Fig. 3-10 *Simplified principle of bipolar feedforward-biased rail-to-rail class-AB output stage.*

V_{REF} should be equal to the supply voltage minus the base-emitter voltages of the output transistors. A similar circuits can be derived in CMOS.

The simplified principle as shown in Fig.3-10 is a very attractive start for creating rail-to-rail feedforward biased output stages. A first possible implementation of this principle is shown in Fig.3-11 in bipolar and CMOS technology[2, 3]. The reference voltage is

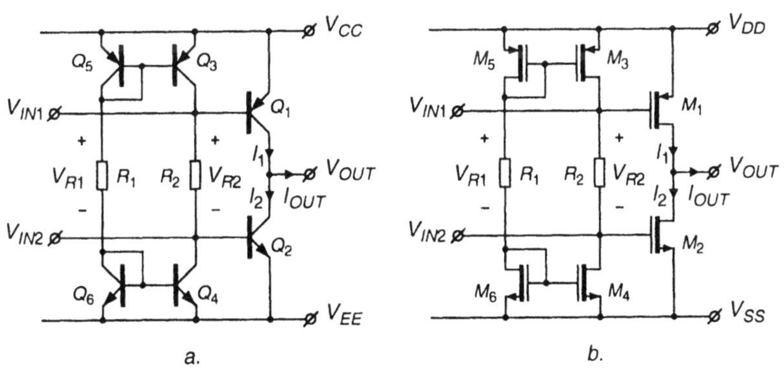

Fig. 3-11 *Rail-to-rail Feedforward-biased resistive-coupled class-AB output stage in bipolar technology (a) and CMOS technology (b).*

realized by resistor R_2. In the bipolar circuit, the bias current for this resistor is generated by connecting two diode connected transistors Q_5 and Q_6, and another resistor R_1

between the supply rails. The bias current is passed to R_2 using current mirrors Q_5, Q_3 and Q_6, Q_4. The relation between the base-emitter voltages V_{BE1} and V_{BE2} of the output stage is given by

$$V_{BE1} + V_{BE2} + V_{R2} = V_{BE5} + V_{BE6} + V_{R1} \tag{3-6}$$

where V_{R1} is the voltage across R_1, and V_{R2} the voltage across R_2. If V_{R1} is equal to V_{R2}, the conventional class-AB behavior described by equation (3-19) is realized. In CMOS technology a similar equation is found yielding the same behavior if the transistors operate in weak inversion. In strong inversion the behavior as described by equation (3-20) is realized. The class-AB characteristics are plotted in Fig.3-12. The asymmetry of the bipolar characteristic is caused by the base current of the PNP transistor. A disadvantage of this circuit is that the quiescent current depends on

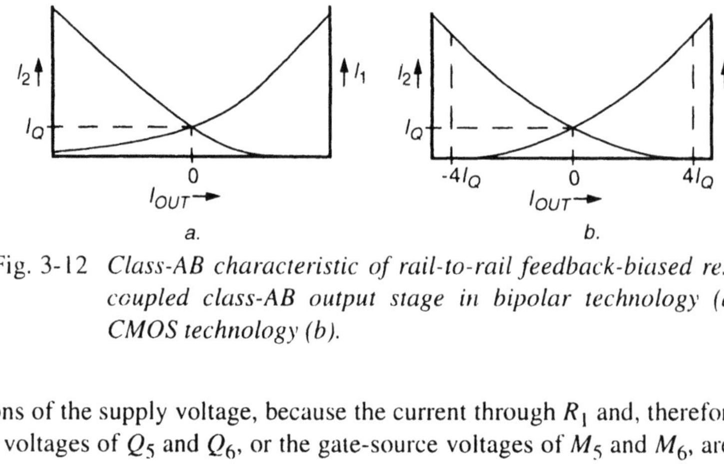

Fig. 3-12 *Class-AB characteristic of rail-to-rail feedback-biased resistive-coupled class-AB output stage in bipolar technology (a) and CMOS technology (b).*

variations of the supply voltage, because the current through R_1 and, therefore, the base-emitter voltages of Q_5 and Q_6, or the gate-source voltages of M_5 and M_6, are influenced by the supply voltage. Auseincr disadvantage is the minimum allowable supply voltage which is determined by the two base-emitter voltages, or by the two gate-source voltages of the output transistors, and is of the order of 1.5 to 1.8 *V* in bipolar technology. The minimum supply voltage in CMOS can be between 1.6 *V* and 2.8 *V*, depending on the process and on the maximum output current. The minimum supply voltage can be reduced by generating a negative voltage across the resistors R_1, R_2 at low supply voltages by adding current sources I_3-I_6, as shown in Fig.3-13 using an example of a CMOS output stage. Thus, the minimum supply voltage can be reduced by an amount equal to the product of the value of the current sources and the value of the resistors.

A remaining disadvantage of the circuit is that a lot of power is consumed by the resistors at higher supply voltages. This can only be improved by removing the direct

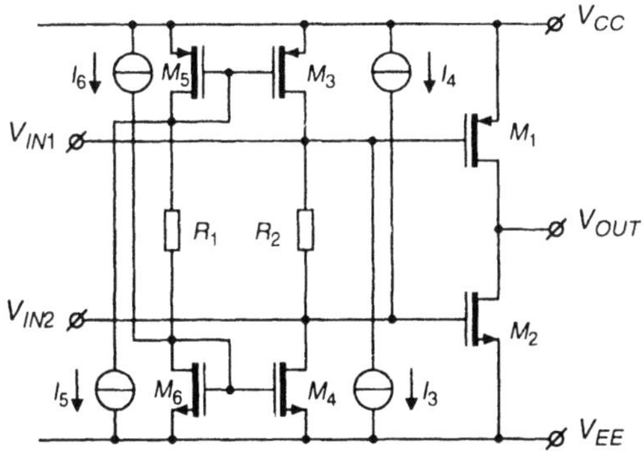

Fig. 3-13 *Resistive-coupled class-AB output stage in CMOS technology which is able to operate at supply voltages below* $2V_{GS}$.

coupling between the bases or gates of the output transistors. A first example of a circuit without direct coupling is shown in Fig.3-14 [4,3]. The output transistors, Q_1 and Q_2,

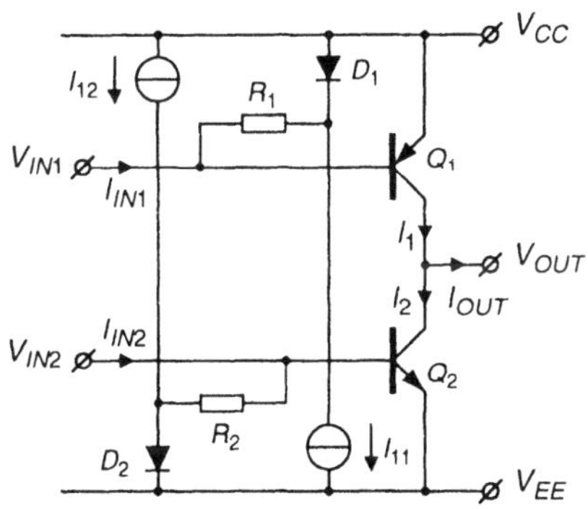

Fig. 3-14 *Separate resistive-coupled bipolar class-AB output stage.*

are biased by the diodes, D_1 and D_2, through the coupling resistors, R_1 and R_2. If the base currents of the output transistors are neglected, the two in-phase input currents, I_{IN1} and I_{IN2}, generate equal voltages across R_1 and R_2, respectively. In such a way, a weak coupling is achieved between the two separate halves of the control circuit so that the sum of the base-emitter voltages of the output transistors is approximately constant. Because the biasing control of the two output transistors is completely separate, the behavior of the output stage relies on the matching of the two input signals I_{IN1} and I_{IN2}. But even if the output stage is driven by two equal input signals the behavior of the circuit is not ideal. If one output transistor is driven hard, the other one is turned off and one of the input signals cannot be used to generate an output signal. Thus, half the input signal is lost and the class-AB biasing is not ideal. The class-AB characteristic is plotted in Fig. 3-15, demonstrating again the asymmetry caused by the base current of the PNP transistor. This circuit can operate on supply voltages as low as 0.9 *V*. Drawbacks are the

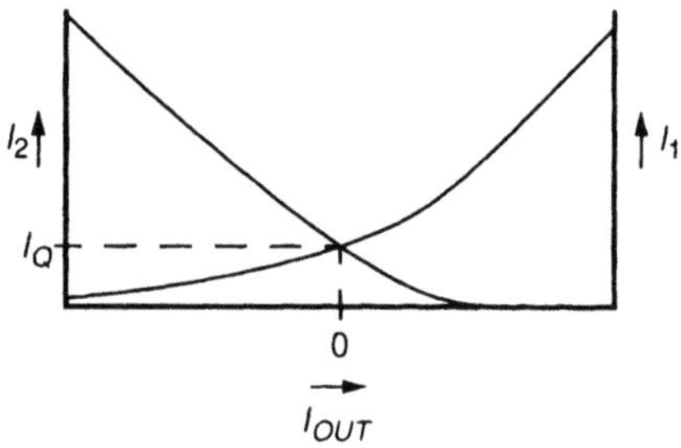

Fig. 3-15 *Class-AB characteristic of separate resistive-coupled bipolar class-AB output stage.*

relatively large loss of signal current in the diodes, and the inaccurate control of the class-AB behavior which is caused by the separate biasing of the output transistors. Due to the loss of signal, the separate resistive coupled technique is not recommended for application in CMOS technology.

The problems of the separate resistive coupled output stage can be solved by preserving the signal current that flows through the resistors R_1 and R_2, and that is lost in diodes D_1 and D_2. We can preserve the signal by inserting cascodes, Q_3 and Q_4, to redirect the signal to the other half of the output stage as shown in Fig. 3-16. Any signal current flowing through resistors R_5, R_6 is now passed by cascodes Q_3, Q_4, respectively, to the other output transistor. Thus, the coupling between the two halves of the control circuit is restored and, a higher gain is obtained because the signal loss is prevented. In this way a mesh is created which generates the strong coupling between the bases of the output transistors. Such a mesh is found in all good class-AB control circuits that do not employ a direct connection between the bases. The mesh establishes the communication between the two separate halves of the control circuit, and redirects signals that are not used to drive the inactive output transistor to the other side where they are used to drive the active output transistor. The class-AB characteristic of the resistive coupled output stage is plotted in Fig. 3-17 where, again, the influence of the base current of the PNP transistors is clearly visible. A disadvantage of the resistive-coupled output stage is the inaccurate biasing caused by the large resistors. If we remove the resistors and use transistors and current sources to realize a similar topology, the circuit shown in Fig. 3-18

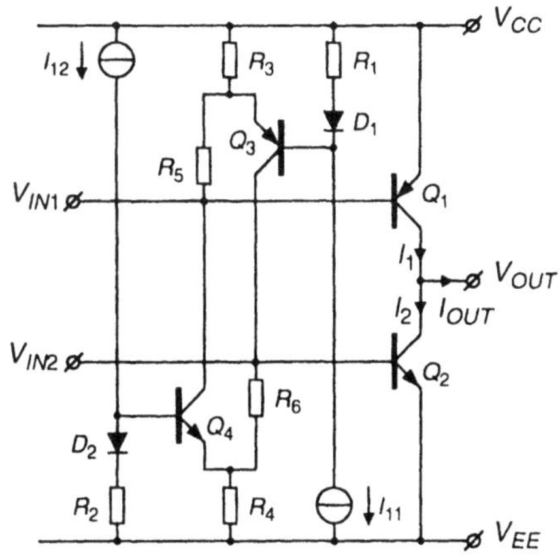

Fig. 3-16 *Resistive-coupled bipolar class-AB output stage with cascodes creating a mesh.*

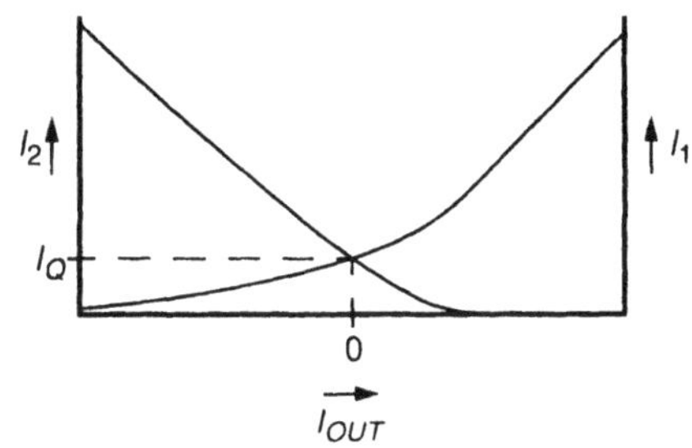

Fig. 3-17 *Class-AB characteristic of resistive-coupled bipolar class-AB output stage.*

is obtained. The biasing of the output transistors, Q_1 and Q_2, is controlled by two translinear loops D_1, Q_3, Q_5, Q_1, and D_2, Q_4, Q_6, Q_2 respectively. The mesh consisting of transistors Q_3-Q_6 which generates the strong coupling between the two translinear control loops is now clearly visible. We call it a folded mesh because it contains a diode levelshift connected to the base of the output transistors that shifts the voltage back to the supply rails. The translinear control loop in combination with the strong coupling of the mesh yields accurate biasing of the output transistors and high gain. As shown in the figure it is possible to drive the mesh by input currents I_{IN1} and I_{IN2} instead of driving the output stage by voltages. In fact, this is possible with all output stages having a similar mesh structure. However, this is not further discussed in this chapter. A problem of the diode-coupled output stage is that the levelshift which is needed in order to correctly bias

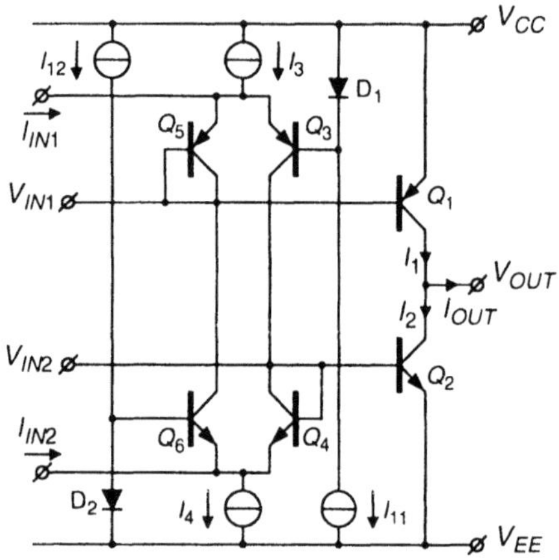

Fig. 3-18 *Diode-coupled bipolar class-AB output stage with folded mesh.*

current sources I_3 and I_4 is difficult to realize. A possible solution is to apply a large scaling between the emitter areas of Q_4 and Q_6, and Q_5 and Q_3. Anyway, there is not much headroom to bias the current sources. Consequently, cascoded current sources cannot be ased. Therefore, the achievable gain of the circuit is limited, especially if CMOS is applied. The class-AB characteristic is depicted in Fig.3-19, displaying fair behavior and, again, the influence of the PNP base current. If we compare the translinear

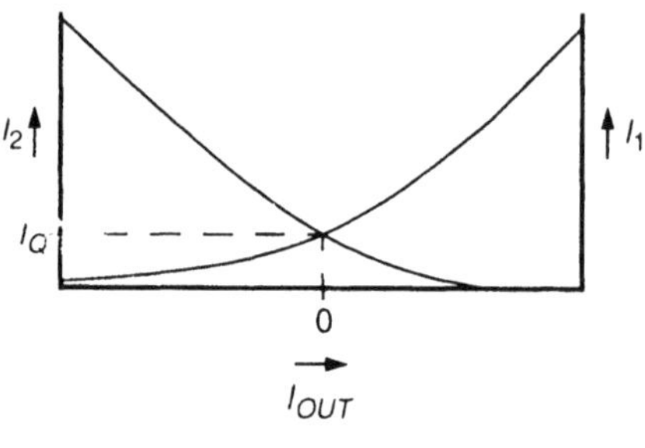

Fig. 3-19 *Class-AB characteristic of diode-coupled bipolar class-AB output stage.*

loops used in this output stage to the translinear loop of a class-AB control with minimum current, as depicted in Fig.3-22, we see that the translinear loops are closely related. In fact, the two transistors Q_3 and Q_6 seem to have the same function as transistors Q_3, Q_4 in Fig.3-22. Therefore, we might wonder why the output stage does not have a class-AB characteristic with minimum current. The answer is found after taking a closer look at the two circuits. All the diodes used in the circuit shown in Fig. 3-22 are biased by a constant current, while in the output stage, as shown in Fig. 3-18, only diodes D_1 and D_2 are biased by a constant current. Diode-connected transistors Q_4 and Q_5, on the other hand, are biased by a current that changes as a function of the output current of the output stage. If, for example, Q_1 delivers a large output current, its base-emitter voltage is large and all the current of current source I_3 and I_4 flow through Q_5 and Q_6. Therefore, the collector current of Q_4 is very small, and the base-emitter voltage of Q_4 is also small. This turns transistor Q_2 off. A minimum current can be maintained in the output stage by either controlling the voltage across Q_4 and Q_5, or by controlling the collector current of Q_4 and Q_5 in such a way that it is constant, or at least remains above a certain minimum value. A solution is, for example, to replace the diode-connected transistors Q_4 and Q_5 by an electronic implementation of a zener diode. A second possibility is depicted in Fig. 3-20, where the improved control circuit is applied in a Darlington output stage consisting of NPN Darlington combination Q_2, Q_{12}, and PNP Darlington combination Q_1, Q_{11}. Additional diode levelshifts D_{21} and

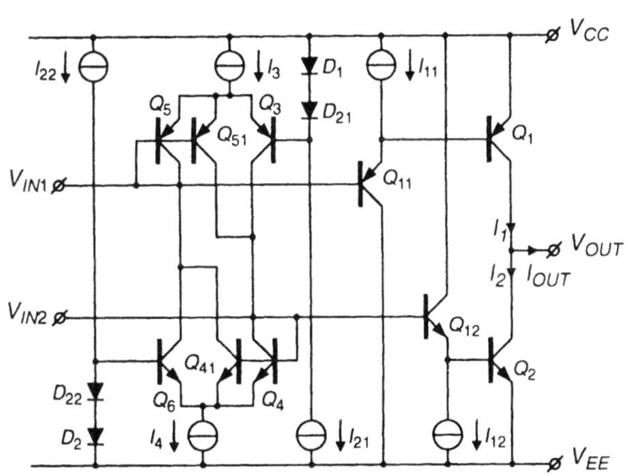

Fig. 3-20 *Diode-coupled class-AB control applied in rail-to-rail Darlington output stage.*

D_{22} have been added to compensate for the additional base-emitter voltage of the Darlington transistors. Due to the levelshift there is sufficient headroom to bias current sources I_3 and I_4. The control circuit is improved by adding two transistors, Q_{41} and Q_{51}, connected in parallel with diode-connected transistors Q_4 and Q_5. The collectors of Q_{41} and Q_{51} are connected to the collectors of Q_6 and Q_3 respectively. This way, if, for example, all the current flows through Q_5 and Q_{51}, some current still flows through Q_4, due to Q_{51}. By straightforward calculations the equation describing the class-AB relation can be found yielding

$$I_1 I_2 - (I_1 + I_2)\frac{1}{3}I_{REF} = \frac{1}{3}I_{REF}^2 \tag{3-7}$$

where it is assumed that transistors Q_{11} and Q_{12} are biased by the same current as the diodes D_{21} and D_{22}, and the emitter areas are equal. After comparison with equation (3-23), which describes class-AB relations in general, we conclude that the quiescent current and minimum current are given by

$$I_Q = I_{REF} \text{ and } I_M = \frac{1}{3}I_{REF} = \frac{1}{3}I_Q \tag{3-8}$$

The influence of the base currents is now negligible because of the Darlington output transistors, as is also clearly demonstrated by the class-AB plot depicted in Fig.3-21 which stays well above the minimum current. Of course, the required supply voltage is

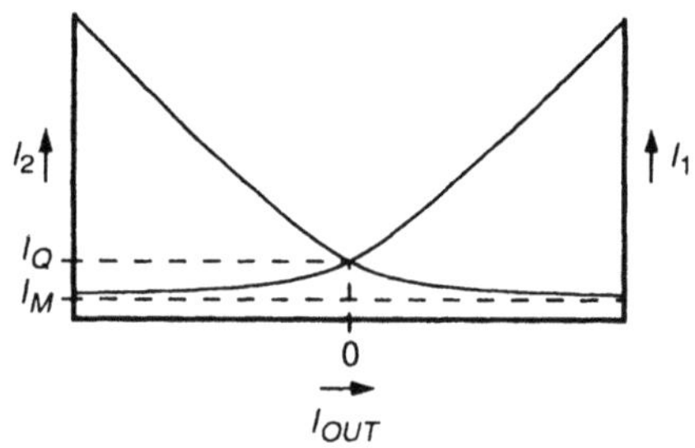

Fig. 3-21 *Class-AB characteristic of diode-coupled rail-to-rail class-AB Darlington output stage.*

increased by a base-emitter voltage and is of the order of 2 V.

At higher supply voltages of the order of 2 *V*, the up-down topology of the translinear loop as used in the previous circuit can be replaced by a standard stacked topology similar to the minimum selector principle discussed in the next section and depicted in Fig. 3-33, yielding a circuit with minimum current equal to half the quiescent current and a simple topology. The resulting control circuit used in a rail-to-rail output stage is

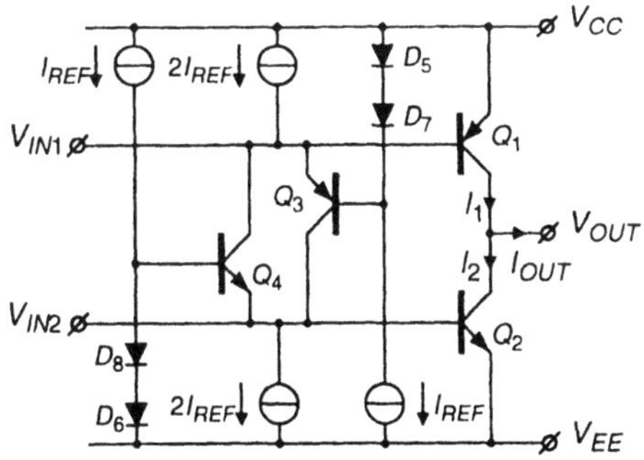

Fig. 3-22 *Bipolar transistor-coupled rail-to-rail class-AB output stage.*

shown in Fig. 3-22. In this circuit the collector current of the output transistors is controlled by two translinear loops Q_2, Q_4, D_6, D_8 and Q_1, Q_3, D_5, D_7[2, 3]. Assuming

$$\frac{A_1 A_3}{A_5 A_7} = \frac{A_2 A_4}{A_6 A_8} \tag{3-9}$$

where A_1-A_4 are the emitter areas of Q_1-Q_4, and A_5-A_8 are the emitter areas of D_5-D_8, the relation between the collector currents of the output transistors is given by

$$I_1 I_2 = (I_1 + I_2)\frac{A_1 A_3}{A_5 A_7}\frac{I_{REF}^2}{I_{REF2}} \tag{3-10}$$

The above assumption as described by equation (3-9) is usually met because it is advantageous to balance the output stage in order to obtain low distortion. The class-AB relation is similar to equation (3-24), which describes minimum selector circuits used in feedback-biased stages, while the quiescent current and minimum current are given by

$$I_Q = 2\frac{A_1 A_3}{A_5 A_7}\frac{I_{REF}^2}{I_{REF2}} \quad \text{and} \quad I_M = \frac{A_1 A_3}{A_5 A_7}\frac{I_{REF}^2}{I_{REF2}} \tag{3-11}$$

The class-AB characteristic is plotted in Fig. 3-23.

The class-AB control circuit contains a mesh consisting of transistors Q_3 and Q_4 which provides a strong coupling between the two inputs, V_{IN1} and V_{IN2}, and, therefore, the output stage can even be driven using only one input. One problem seems to be that the

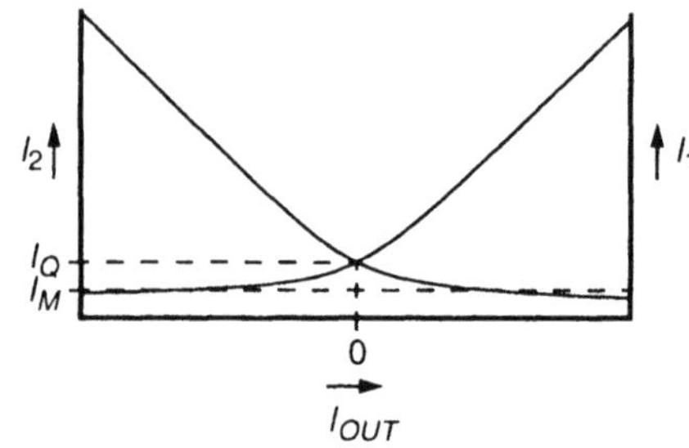

Fig. 3-23 *Class-AB characteristic of bipolar transistor-coupled rail-to-rail class-AB output stage.*

impedance at the inputs V_{IN1} and V_{IN2} of the output stage is reduced by the emitters of Q_3 and Q_4. However, due to the positive feedback in the mesh Q_3, Q_4, any signal current flowing into the emitters will return. The signal just cannot be lost. This is a property shared by all class-AB control circuits having a similar mesh structure, such as the diode-connected control circuits. However, an additional advantage over low-voltage diode-connected output stage is that there is now sufficient room to bias current sources I_3 and I_4 in such a way that their output impedance is very high. Thus, the input impedance of the output stage is not degraded by the class-AB control circuit, so that this configuration is especially interesting in CMOS, as shown in Fig. 3-24 [5]. In CMOS

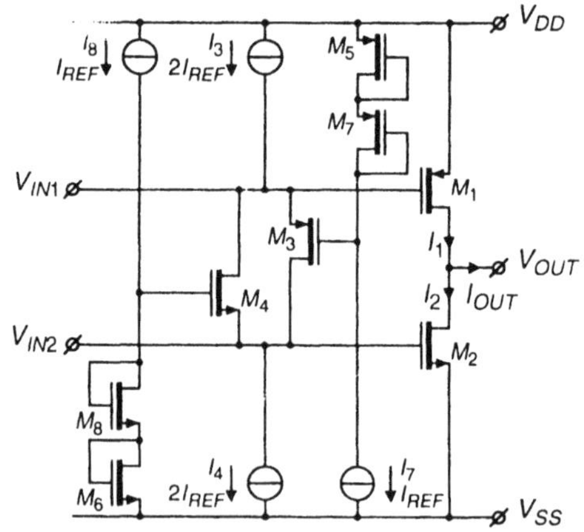

Fig. 3-24 *CMOS transistor-coupled rail-to-rail class-AB output stage*

technology the signal cannot be lost by flowing into the class-AB control transistors since the mesh from node V_{IN1} via M_3 to V_{IN2}, and back again from V_{IN2} to V_{IN1} via M_4 is only connected to high impedances, the gates of M_1-M_4, and the impedance of the current sources I_3 and I_4. An alternative way to understand the signal behavior of the circuit is by calculating the impedance of the mesh, which should be sufficiently low in order to obtain a strong coupling between the two inputs. Therefore, we calculate the impedance $z_{in1,in2}$ between the gates of M_1 and M_2. The impedance is found by dividing the signal voltage v_{in1}-v_{in2} across the gates of M_1 and M_2 by the current flowing through the mesh transistors yielding

$$z_{in1,\,in2} = \frac{v_{in1} - v_{in2}}{g_{m3}v_{in1} - g_{m4}v_{in2}} \qquad (3\text{-}12)$$

where g_{m3} is the transconductance of M_3, g_{m4} the transconductance of M_4, v_{in1} the voltage at the gate of M_1, and v_{in2} the voltage at the gate of M_2. In the quiescent state equation (3-12) reduces to

$$z_{in1,\,in2} = \frac{1}{g_{m3}} \qquad (3\text{-}13)$$

where it is assumed that g_{m3} and g_{m4} are approximately equal. For balancing reasons, M_{13} and M_{14} are usually designed in such a way that their transconductance is equal in the quiescent state. As we see the impedance of the mesh is low, certainly compared to the high output impedances of current source I_3 and I_4. As a result, the gates of M_1 and M_2 are practically connected in parallel by the operation of the class-AB control. If, on the other hand, one output transistor, M_1 for example , is handling a large output current, the other output transistor M_2 is regulated at a constant minimum current. Coupling of the gates of the two output transistors using a mesh with low impedance is not possible because M_1 handles all the signal and has a large signal swing at its gate, while M_2 has a constant gate-source voltage. Consequently, the small-signal voltage v_{in2} at the gate of M_2 is very small. Because the gate-source voltage of M_1 is large, M_3 is off and its transconductance is very small. Using equation (3-12) it can be concluded that the mesh impedance is very large so that there is no coupling between the two gates of the output transistors. Therefore, M_1 can deliver large output currents without being limited by the inactive output transistor. The signal at the gate of M_2 is directed to the active output transistor M_1 by cascode M_4. Similar results are found for the bipolar version.

In CMOS the relation between the drain currents I_1 and I_2 is similar to equation (3-28), which describes CMOS minimum selector circuits used in feedback-biased stages. The relation of the feedforward-biased stage is given by [6]

$$\left(\sqrt{I_1}-\alpha\sqrt{I_q}\right)^2+\left(\sqrt{I_2}-\alpha\sqrt{I_q}\right)^2 = 2(\alpha-1)^2 I_q \qquad (3\text{-}14)$$

where it is assumed that the transistors operate in strong inversion, and that

$$\frac{\left(\frac{W}{L}\right)_1}{\left(\frac{W}{L}\right)_2} = \frac{\left(\frac{W}{L}\right)_3}{\left(\frac{W}{L}\right)_4} = \frac{\left(\frac{W}{L}\right)_5}{\left(\frac{W}{L}\right)_6} = \frac{\left(\frac{W}{L}\right)_7}{\left(\frac{W}{L}\right)_8} \quad \text{and} \quad \alpha = 1+\sqrt{\left(\frac{W}{L}\right)_6\left(\frac{L}{W}\right)_8} \qquad (3\text{-}15)$$

Further, it is assumed that M_1 and M_5, as well as M_2 and M_6, have the same gate-source voltage in order to compensate for the body effect. The drain currents of the output transistors obey equation (3-14) until one of them exceeds a value of $\alpha^2 I_Q$. At that moment one of the transistors M_4, M_8 is cut off and the drain current of the other output transistor stays at a minimum current, given by

$$I_M = \left(\alpha-\sqrt{2}(\alpha-1)\right)^2 I_Q \qquad (3\text{-}16)$$

If M_6 and M_8 have the same sizes, the minimum current is about $0.34I_Q$. The class-AB relation is plotted in Fig.3-25. A disadvantage of the circuit shown in Fig.3-24 is that

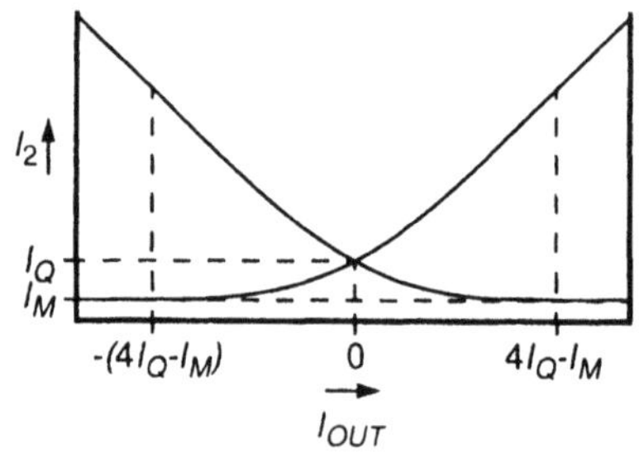

Fig. 3-25 *Class-AB characteristic of bipolar transistor-coupled rail-to-rail class-AB output stage.*

two stacked diodes are used which allows the circuit only to operate on supply voltages down to 1.8 V in bipolar technology, and between 1.6 V and 2.8 V in CMOS technology depending on the process and the maximum output current.

3.5 Feedback class-AB biasing

Using feedforward class-AB control as discussed in the previous section elegant class-AB output stages can be obtained, but proper class-AB control is not possible at very low supply voltages. Accurate control of the class-AB biasing of the output transistors at very low supply voltages is only possible using feedback biasing [7,8,9]. A generalized way to control the biasing of the output transistors using a feedback loop is depicted in Fig.3-26. The collector currents of the output transistors are measured using transistors

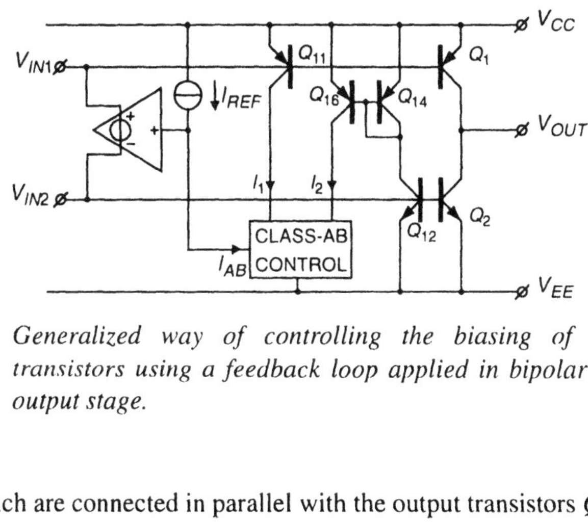

Fig. 3-26 *Generalized way of controlling the biasing of the output transistors using a feedback loop applied in bipolar rail-to-rail output stage.*

Q_{11} and Q_{12} which are connected in parallel with the output transistors Q_1 and Q_2. Thus, the collector current of Q_{11} is proportional to the collector current of Q_1, and the collector of Q_{12} is proportional to the collector current of Q_2. The collector current of I_{11} is directly used as input I_1 of the class-AB control block. The collector current of I_{12} is mirrored by current mirror Q_{14}, Q_{16}, and used as input I_2 by the class-AB control block. The control block contains a circuit that generates a current I_{AB} that depends on the currents I_1 and I_2. The current I_{AB} is compared to a reference I_{REF} and the difference is amplified by the amplifier which control the inputs of the output transistors and completes the feedback loop. The feedback loop forces the current I_{AB} to be equal to the reference current I_{REF}.

Advantages of the feedback biasing techniques are the freedom of design and the high accuracy of the class-AB control implementation. The design freedom is the result of the separation of the measuring of the output-stage bias currents and the implementation of the class-AB control law. This enables the design of reliable circuits that can operate on the lowest possible supply voltages. Establishing the desired class-AB behavior is easy since since any type of class-AB behavior can be realized by replacing the class-AB control block by a suitable circuit. The high accuracy can be simply obtained by creating high gain in the class-AB control loop. The well-known consequence of applying feedback, however, is that the circuit may oscillate and therefore the stability has to be taken into account. We first focus on the implementation of the class-AB control block. Afterwards the properties of the feedback loop are investigated.

In order to obtain straightforward and elegant class-AB control circuits, the class-AB relation should be simple. Therefore, we start by examining the simplest possible equation, a linear relation between the two bias currents of the output transistors as given by

$$I_1 + I_2 = 2I_Q \tag{3-17}$$

This relation is shown in Fig.3-27 in the plot on the left-hand side. For currents smaller than $2I_Q$, both transistors are biased and the output stage operates in class-A mode. As soon as the bias current of one of the transistors exceeds the value $2I_Q$, the other transistor is completely cut off having zero bias current. The stage is then operating under a class-B regime and the behavior is no longer described by equation (3-17). The currents as a function of the output current I_{OUT} are shown in Fig.3-27 on the right-hand side. As we see a control based on equation (3-17) is sufficient to set the quiescent

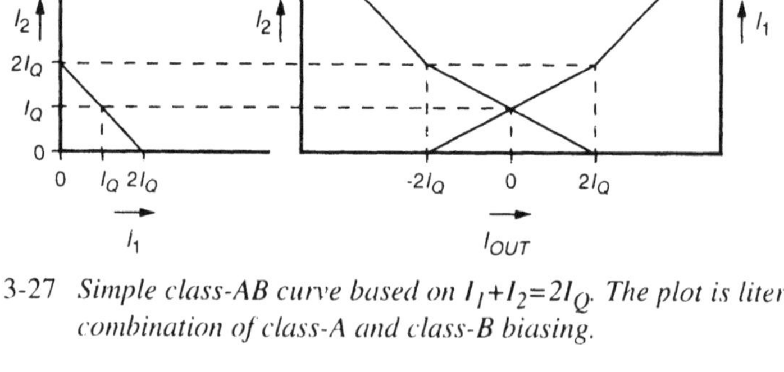

Fig. 3-27 *Simple class-AB curve based on $I_1+I_2=2I_Q$. The plot is literally a combination of class-A and class-B biasing.*

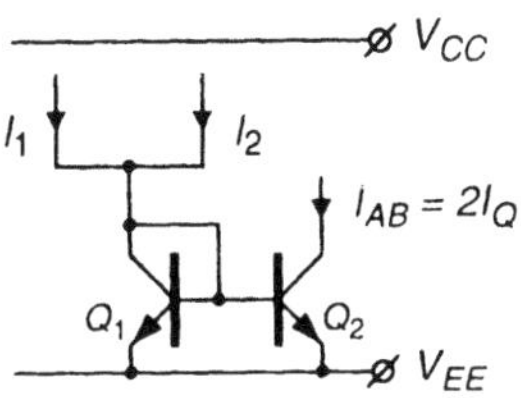

Fig. 3-28 *Simple implementation of linear current class-AB control described by* $I_1+I_2=2I_Q$.

current but cut off of the output transistors cannot be prevented. However, the relation can be simply implemented using a current-summing node as shown in Fig. 3-28. Using the current mirror Q_1, Q_2, the circuit can be directly inserted into the class-AB control block of the circuit shown in Fig. 3-26, yielding a quiescent current I_Q that is equal to half the value of I_{REF}.

Alternatively, we can consider a linear relation consisting of base-emitter voltages or gate-source voltages described by

$$V_1 + V_2 = 2V_Q \tag{3-18}$$

Such a relation can be simply implemented using a mesh consisting of base-emitter voltages or gate-source voltages as shown in Fig. 3-29, using bipolar technology which can directly be inserted into the class-AB control block of the circuit shown in Fig. 3-26, resulting in a quiescent current that is equal to I_{REF}. Such a mesh is known as a translinear loop, or MOS translinear loop when MOS technology is used. Due to the non-

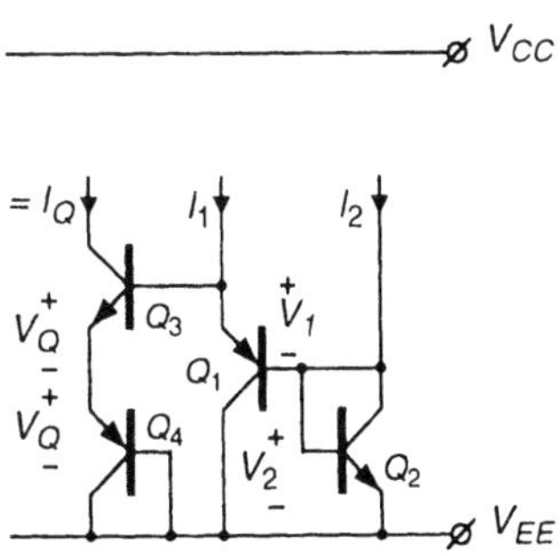

Fig. 3-29 *Simple implementation of linear voltage class-AB control described by* $V_1+V_2=2V_Q$ *using a mesh of base-emitter voltages or translinear loop.*

linear relation between voltage and current found in transistors, the resulting relation between the currents is much more gradual. In bipolar technology and in CMOS technology operating in weak inversion, the sum of voltages leads to a product of currents yielding the conventional class-AB relation as given by

$$I_1 I_2 = I_Q^2 \tag{3-19}$$

The relation between I_1 and I_2 and the class-AB characteristic as a function of the output current are depicted in equation (3-19). Clearly, the class-AB plot is much more gradual and the currents never reach zero. Therefore, theoretically, cut-off of the output

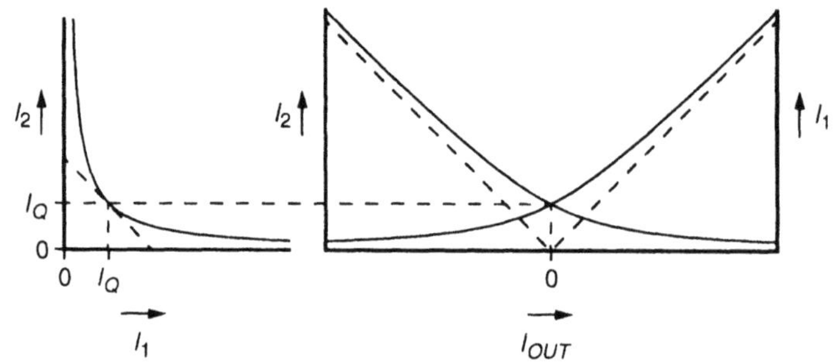

Fig. 3-30 *Conventional class-AB curve based on $V_1+V_2=2V_Q$ using bipolar technology or CMOS technology in weak inversion yielding $I_1\, I_2=I_Q^2$.*

transistors is prevented when the conventional class-AB control is used to bias the output stage. In practical cases, however, non idealities such as base currents cause cut-off if one of the output transistors is driven hard. In CMOS in strong inversion the relation between the currents is given by

$$\sqrt{I_1} + \sqrt{I_2} = 2\sqrt{I_Q} \tag{3-20}$$

which is only valid for currents smaller than $4I_Q$. As can be concluded from the class-AB plots shown in Fig. 3-31 cut-off occurs when the bias current of one of the transistors exceeds the value of $4I_Q$. Although a more gradual behavior is obtained than with the linear class-AB relation, cut-off is not prevented using a CMOS version of the conventional class-AB control law in strong inversion. It should be noted that before one of the currents reaches the value of zero, the corresponding MOS transistor enters the weak inversion region. Therefore, cut-off of that transistor is slightly delayed. However,

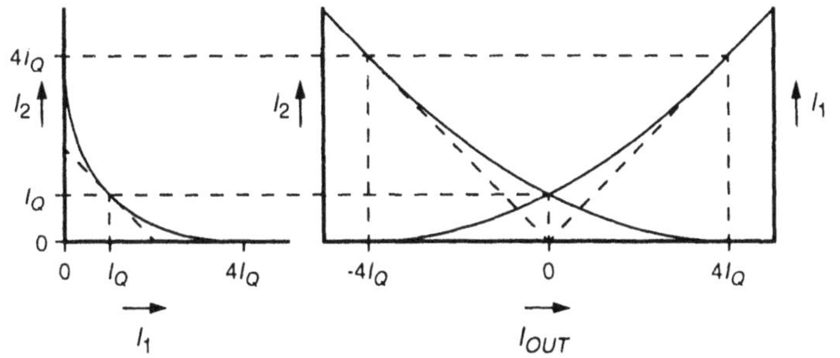

Fig. 3-31 *Conventional class-AB curve based on $V_1+V_2=2V_Q$ using CMOS technology in strong inversion yielding $\sqrt{I_1}+\sqrt{I_2}=2\sqrt{I_Q}$.*

the active transistor operates in strong inversion with a gate-source voltage proportional to the square root of its drain current. In order to obey the sum-of-gate-source-voltages control law therefore, the gate-source voltage of the inactive transistor quickly diminishes to zero when the active transistor is driven hard. In conventional common-drain class-AB output stages, the gate-source voltage of the inactive transistor may even become negative. Although the switching of MOS transistors gives less problems than the switching of bipolar transistors, it may take some time before the gate-source voltage of the output transistor is slewed to the required value. Thus, both in bipolar and in CMOS technology, a class-AB control without cut-off is preferable in order to obtain low distortion. Therefore, class-AB control architectures have been invented that, in addition to controlling the quiescent current, also control a minimum current. The conventional class-AB characteristic depicted in Fig. 3-30 already displays quite satisfactory behavior and is therefore taken as a starting point to derive minimum-current based control laws. Using the conventional characteristic, a control with minimum

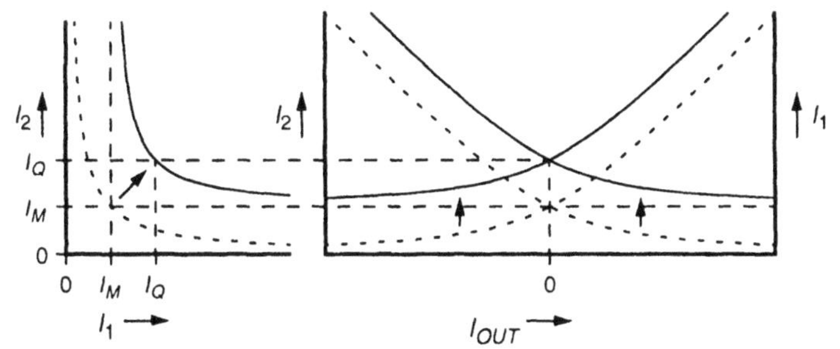

Fig. 3-32 *Translation of conventional class-AB curve in order to obtain a class-AB characteristic with minimum current I_M.*

current can be obtained by applying a translation as depicted in Fig. 3-32. The conventional characteristic is plotted using dotted lines. The translation is described by

$$I_1 = I'_1 - I_M \qquad I_2 = I'_2 - I_M \qquad I_Q = I'_Q - I_M \tag{3-21}$$

where I_M is the minimum current and the primed variables indicate the new variables. Applying the translation to equation (3-19) gives

$$(I'_1 - I_M)(I'_2 - I_M) = (I'_Q - I_M)^2 \tag{3–22}$$

After rearranging terms and removing the primes, the minimum-current control law is found as

$$I_1 I_2 - (I_1 + I_2) I_M = I_Q (I_Q - 2I_M) \tag{3-23}$$

which describes a minimum-current law where the minimum current can be chosen arbitrarily. The equation, however, is complicated, leading to a complex control circuit. Fortunately, by choosing a clever value for the minimum current the equation can be simplified. By taking I_M equal to half the quiescent current I_Q the right-hand side of equation (3-23) vanishes yielding

$$I_1 I_2 = (I_1 + I_2) I_M \qquad \text{with} \qquad I_M = \frac{1}{2} I_Q \tag{3-24}$$

The relation between I_1 and I_2 and the resulting class-AB plot is shown by solid lines as depicted in Fig.3-32. The control law controls the minimum current as well as the quiescent current, while the relation is only slightly more complex than the conventional relation. It combines smooth switch over with a relatively high minimum current as required for obtaining low distortion class-AB control. After rewriting equation (3-24) we find

$$\frac{I_M}{I_1} I_M + \frac{I_M}{I_2} I_M = I_M \tag{3-25}$$

which can easily be implemented by using two translinear loops to generate currents proportional to I_M^2/I_1 and I_M^2/I_2 and a current-summing node to sum these currents. An example is depicted in Fig. 3-33. The circuit can be immediately used in the output stage depicted in Fig. 3-26, which results in a minimum current that is equal to I_{REF}. The first translinear loop consists of Q_1, Q_3, Q_5, and Q_6 which forces the collector current in Q_3 to be equal to I_M^2/I_1. The second loop consisting of Q_2, Q_4, Q_5, and Q_6 generates a collector current of Q_4 which equals I_M^2/I_2. By using the emitters of two PNP transistors as the current-summing node, a differential pair Q_3, Q_4 results. The output voltage at the emitters of the differential pair is controlled by the smaller of the two voltages at the bases of Q_3 and Q_4 which corresponds to the smaller of the two input currents I_1, I_2. The output current I_M is determined by the smaller of the two input currents and, therefore, this circuit is usually called a minimum selector. The circuit is very useful and forms the basis of an important class of feedback biased output stages. The minimum required supply voltage can be very low if transistors Q_1 and Q_2 are replaced by resistors. The

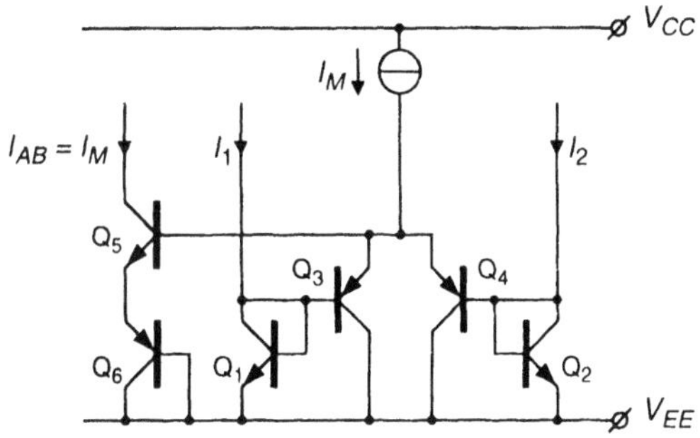

Fig. 3-33 *Compact implementation of class-AB characteristic with minimum current based on equation (3-25) using a decision pair.*

same circuit can also be applied in CMOS technology as shown in Fig. 3-34. When the

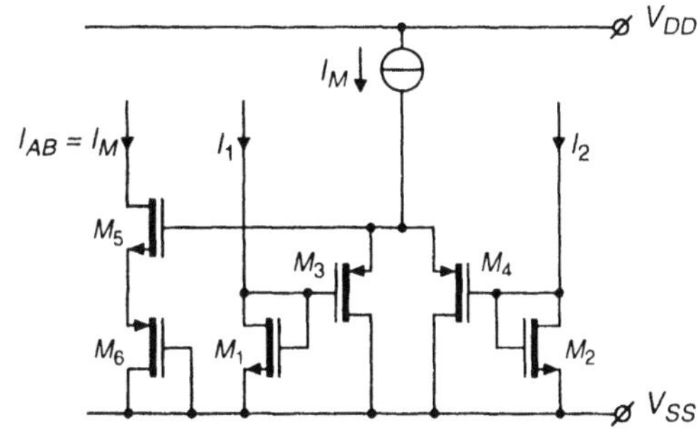

Fig. 3-34 *CMOS implementation of compact class-AB control with minimum current by using a decision pair.*

transistors operate in weak inversion, exactly the same equations and the same characteristic is implemented as by the bipolar version. However, in strong inversion different equations are found due to the quadratic relation between drain current and gate-source voltage. Assuming the *W/L* ratios and threshold voltages of all matching transistors are equal, the drain current I_3 of M_3 in strong inversion is given by

$$\sqrt{I_3} = 2\sqrt{I_M} - \sqrt{I_1} \tag{3-26}$$

and the drain current I_4 of M_4 by

$$\sqrt{I_4} = 2\sqrt{I_M} - \sqrt{I_2} \tag{3-27}$$

Combining equation (3-26) and equation (3-27) with the nodal equation of the summing node, a relation is obtained which describes the class-AB characteristic of the MOS version as given by

$$\left(2\sqrt{I_M} - \sqrt{I_1}\right)^2 + \left(2\sqrt{I_M} - \sqrt{I_2}\right)^2 = I_M \tag{3-28}$$

If we plot $\sqrt{I_1}$ as a function of $\sqrt{I_2}$, equation (3-28) describes a circle with center ($2\sqrt{I_M}$, $2\sqrt{I_M}$) and radius $\sqrt{I_M}$ [10]. Further, it is important to note that the above equations are only valid for values of I_1 and I_2 smaller than $4I_M$. If I_1 or I_2 is larger than $4I_M$, transistor M_3 or M_4, respectively, is off and the other current I_2 or I_1, respectively, remains constant

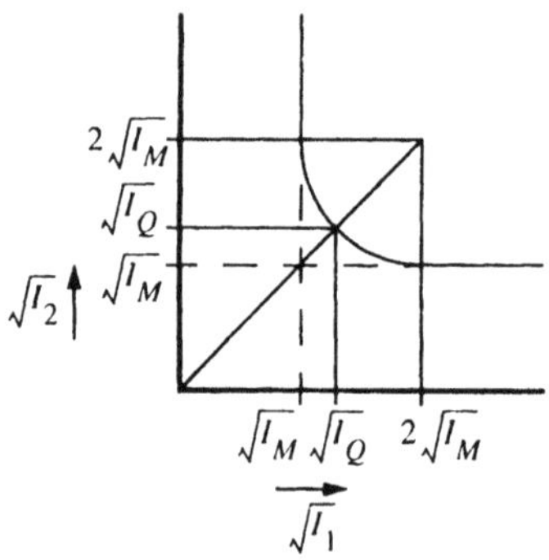

Fig. 3-35 *Plot of $\sqrt{I_1}$ as a function of $\sqrt{I_2}$ describing the CMOS class-AB characteristic with minimum current as given by equation (3-28).*

and is equal to I_M. The resulting plot of $\sqrt{I_1}$ as a function of $\sqrt{I_2}$ is shown in Fig. 3-35. By inspection of the graph it is easy to find the relation between I_M and I_Q as given by

$$I_Q = \left(\frac{2\sqrt{2}-1}{\sqrt{2}}\right)^2 I_M \approx 1.67 I_M \tag{3-29}$$

However, in practical circuits different values may be found, depending on W/L ratios and the ratio between the bias currents of M_5, M_6 and the bias current of M_3, M_4. A final minimum selector circuit of interest is based on another version of equation (3-24) as given by

$$I_1 (I_2 - I_M) = I_2 I_M \tag{3-30}$$

The above relation is enforced by the CMOS circuit as shown in Fig. 3-36, where all transistors operate in weak inversion. It can be applied in a CMOS version of the circuit

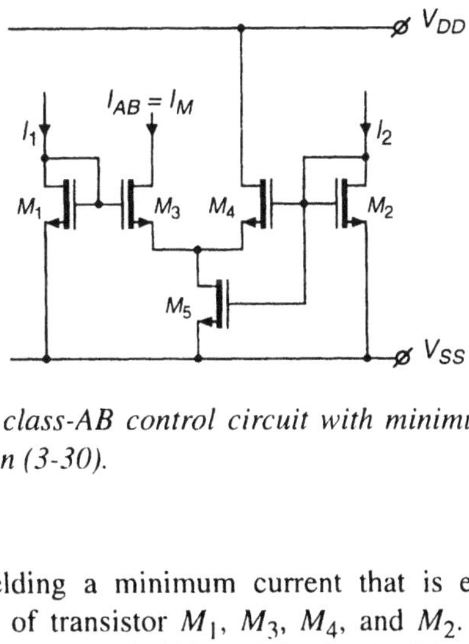

Fig. 3-36 *CMOS class-AB control circuit with minimum current based on equation (3-30).*

shown in Fig. 3-26 yielding a minimum current that is equal to I_{REF}. The MOS translinear loop consists of transistor M_1, M_3, M_4, and M_2. Transistor M_5 supplies a current equal to I_2 to the sources of differential pair M_3, M_4 so that the drain current of M_4 is equal to I_2-I_M. The circuit can be further simplified by using a MOS transistor operating in the linear region. It is easy to show that the relation describing the operation of a MOS transistor operating in the linear region and in strong inversion can be decomposed as [11]

$$
\begin{aligned}
I_D &= \mu_n C_{ox} \frac{W}{L} V_{DS}\left(V_{GS} - V_{TH} - \frac{1}{2} V_{DS} \right) \\
&= \frac{1}{2}\mu_n C_{ox} \frac{W}{L} (V_{GS} - V_{TH})^2 - \frac{1}{2}\mu_n C_{ox} \frac{W}{L} (V_{GD} - V_{TH})^2
\end{aligned}
\tag{3-31}
$$

The right-hand side of equation (3-31) describes two saturated MOS transistors connected in anti-parallel as depicted Fig. 3-37a. This property is also found with bipolar transistors as expressed by the Ebers-Moll model which consists of two anti-parallel connected transistors which are both active when the device is saturated. Since it does not matter for signals if the drain of M_4 is connected to V_{SS} or V_{DD}, the combination of M_4 and M_5 in Fig. 3-36 is identical to the circuit shown in Fig. 3-37a. Thus, M_4 and M_5 can be replaced by a single MOS transistor operating in the linear mode, yielding the simple minimum selector circuit as depicted in Fig. 3-37b. The circuit is very compact

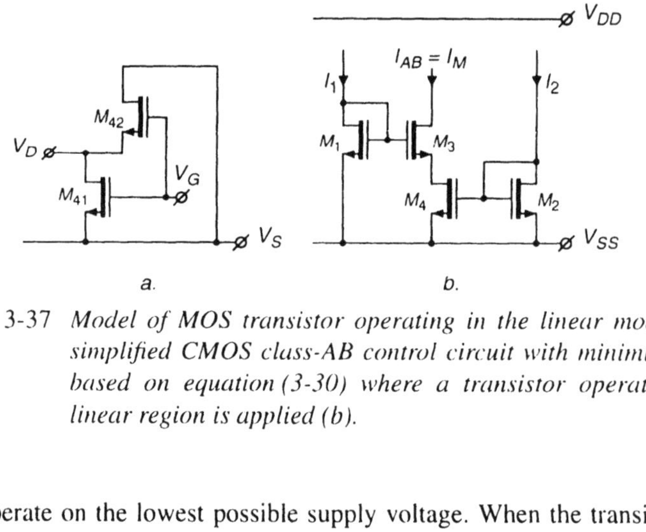

Fig. 3-37 *Model of MOS transistor operating in the linear mode (a) and simplified CMOS class-AB control circuit with minimum current based on equation (3-30) where a transistor operating in the linear region is applied (b).*

and can operate on the lowest possible supply voltage. When the transistors operate in weak inversion, the behavior as depicted in Fig. 3-32 is implemented and the class-AB characteristic is described by equation (3-30) [11]. In strong inversion the behavior is slightly different. The transfer function from the drain current of M_1 to the drain current of M_3 can be expressed as a function of the drain-source voltage of M_4

$$\sqrt{\frac{2I_1L_1}{\mu_nC_{ox}W_1}} = \sqrt{\frac{2I_ML_3}{\mu_nC_{ox}W_3}} + V_{DS4} \qquad (3\text{-}32)$$

where it is assumed that the threshold voltages of M_1 and M_3 are equal. Then the expression for M_4 in the linear region is combined with the expression for the gate-source voltage of M_2, yielding a relation between V_{DS4}, I_2 and I_M

$$I_M = \mu_nC_{ox}\frac{W_4}{L_4}\left(V_{DS4}\sqrt{\frac{2I_2L_2}{\mu_nC_{ox}W_2}} - \frac{1}{2}V_{DS4}^2\right) \qquad (3\text{-}33)$$

where it is assumed that the threshold voltages of M_1 and M_3 are equal. Further, we assume that the length L of all transistors is equal. Substituting the value of V_{DS4} which results from equation (3-32) in equation (3-33) and assuming the width W of all transistors is equal yields

$$I_2 = \left(\frac{I_M - \sqrt{I_1}\left(\sqrt{I_M} - \sqrt{\frac{1}{4}I_1}\right)}{\sqrt{I_1} - \sqrt{I_M}}\right)^2 \qquad (3\text{-}34)$$

It should be noted that the above equations are only valid if M_4 is operating in the linear region. The edge of the linear region is reached for $I_1 = 4I_M$ if all transistor widths are equal. For larger values of I_2, M_4 enters the saturation mode and I_2 is constant and equal to I_M. From equation (3-34) it can be concluded that there is an asymptote for $I_1 = I_M$. Another interesting point is found when I_1 is equal to I_2. Then, both currents are equal to $2I_M$. By using the model shown in Fig. 3-37a, a relation between I_1 and I_2 can be derived much faster. If the W over L ratio of all transistors is equal, the drain current of M_{41} is equal to I_2, and the drain current of M_{42} is equal to I_2-I_M yielding [12]

$$I_1 = \left(\sqrt{I_2} + \sqrt{I_M} - \sqrt{I_2 - I_M}\right)^2 \qquad (3\text{-}35)$$

Although equation (3-34) and equation (3-35) look completely different they are identical which can be easily proven by substituting equation (3-35) in equation (3-34). The relation is plotted in Fig. 3-38 on the left-hand side together with the plot of the characteristic when the transistors operate in weak inversion which is indicated by dashed lines. The resulting class-AB plot is shown on the right-hand side. The class-AB curve in strong inversion reaches an asymptote at

$$I_{OUT} = I_1 - I_2 = I_M - 4I_M = -3I_M \qquad (3\text{-}36)$$

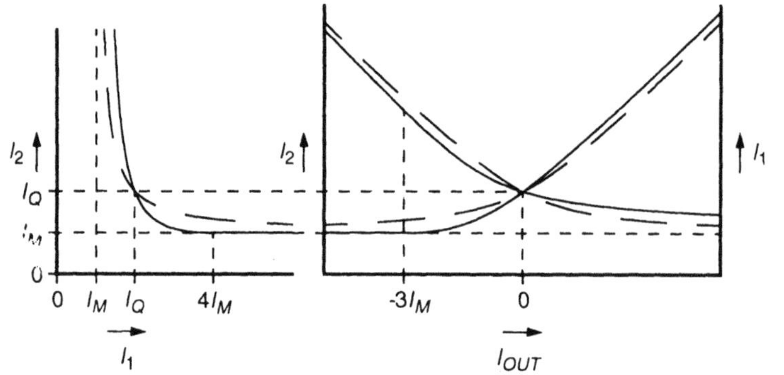

Fig. 3-38 *Class-AB characteristic of CMOS compact minimum selector circuit. The characteristic in strong inversion is indicated by a solid line, while the characteristic in weak inversion is indicated by a dashed line.*

We have discussed several techniques to implement a suitable class-AB relation. Now it is time to return to the main properties of class-AB feedback loops. To investigate some general properties regarding the stability of class-AB feedback loops the circuit shown in Fig. 3-39 is used. Measuring transistors M_{11} and M_{12} are connected in parallel with the output transistors M_1 and M_2, so that the drain currents I_{11} and I_{12} of M_{11} and M_{12} are proportional to the drain currents I_1 and I_2 of the output transistors. The currents are fed into a class-AB control block which contains one of the previously discussed translinear circuits in order to establish a desired class-AB characteristic. The output voltage V_{AB} of the control block depends on the input currents of the block I_{11} and I_{12} in such a way that the desired class-AB equation is satisfied. Using a differential amplifier M_{13}, M_{14}, V_{AB} is compared to reference voltage V_{REF}. The difference is amplified by the differential amplifier which controls the gates of M_{11} and M_{12} and thereby completes the feedback loop. The feedback loop regulates V_{AB} equal to V_{REF} fixing the desired class-AB behavior. Frequency compensation of the output stage is achieved using Miller capacitors C_{M1A} and C_{M1B}. Further, the gate-source capacitances of M_1, M_{11} and M_2, M_{12} are modelled by C_{GS1} and C_{GS2} respectively. The class-AB loop transfer function can be evaluated, for instance, by breaking the loop between the gate of M_{14} and the output of the class-AB control block, and calculating the transfer function from the

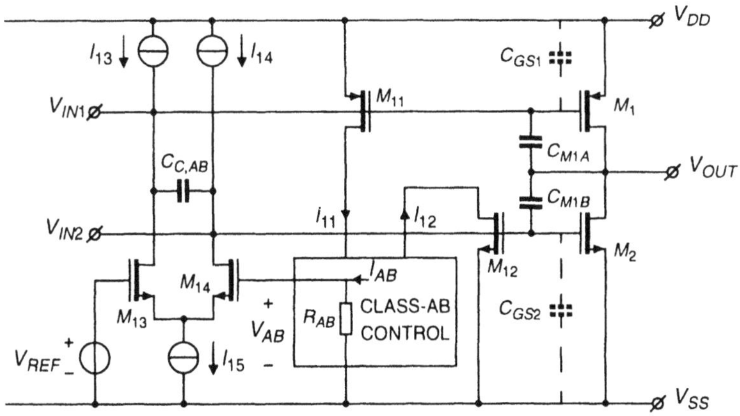

Fig. 3-39 *Feedback-biased CMOS class-AB output stage with differential control amplifier.*

voltage at the gate of M_{14} to the voltage V_{AB}. Also, it is important to note that the node V_{OUT} can be considered as ground for class-AB signals since the overall feedback around the amplifier forces the output voltage only to respond to input signals of the main amplifier. In order to evaluate the loop gain we need the small signal transfer function of the class-AB control block. The transfer function of the control block depends on the actual class-AB characteristic. In general, we can express the small signal output voltage of the control block by

$$v_{AB} = i_{AB} r_{AB} = r_{AB}(k_1 i_{11} + k_2 i_{12}) \tag{3-37}$$

where i_{AB} is the signal current corresponding to the large-signal output current of the control block I_{AB}, r_{AB} is the small signal resistance corresponding to the output impedance of the control block R_{AB}, and i_{11} and i_{12} are small-signal currents corresponding to the large-signal input currents of the control block I_{11} and I_{12} respectively. The variables k_1 and k_2 depend on the actual state of the output stage, or the ratio of i_{11} and i_{12}, and the applied class-AB relation. We can now continue to investigate the class-AB loop.

Basically, the class-AB loop can be considered as a single stage amplifier with transconductance stage M_{13}, M_{14} and with the Miller capacitors and the gate-source capacitors as load capacitors. The unity-gain frequency of the class-AB loop is therefore given by

$$\omega_{0,AB} = g_{AB} r_{AB} \cdot \left(\frac{k_1 g_{m11}}{C_{M1A} + C_{GS1}} + \frac{k_2 g_{m12}}{C_{M1B} + C_{GS2}} \right) \tag{3-38}$$

where g_{AB} is the transconductance of class-AB amplifier M_{13}, M_{14}, g_{m11} is the transconductance of transistor M_{11} and g_{m12} is the transconductance of transistor M_{12}. In order to find the variables k_1 and k_2, we consider the following three examples of class-AB curves: the conventional class-AB relation in bipolar technology; the conventional class-AB relation in CMOS technology, with transistors operating in strong inversion; and the bipolar class-AB relation with minimum-current control. In order to implement the bipolar conventional class-AB characteristic as described by equation (3-19), the large signal transfer function from I_{11} and I_{12} to I_{AB} should be

$$I_{11} I_{12} = I_{AB}^2 \tag{3-39}$$

from which the small signal transfer function can be derived as given by

$$i_{AB} = \frac{1}{2}\frac{I_2}{I_Q} i_{11} + \frac{1}{2}\frac{I_1}{I_Q} i_{12} \quad \text{or} \quad i_{AB} = \frac{1}{2}\sqrt{\frac{I_2}{I_1}} i_{11} + \frac{1}{2}\sqrt{\frac{I_1}{I_2}} i_{12} \tag{3-40}$$

where it has been used that the ratio between the W/L ratios of M_{11} and M_1 is equal to the ratio between the W/L ratios of M_{12} and M_2 for reasons of distortion, and the parameters I_{11}, I_{12}, and I_{AB} can be replaced by the output stage parameters I_1, I_2, and I_Q respectively. equation (3-40) describes the transfer function of the class-AB control of the conventional output stage in bipolar technology and in CMOS technology with transistors operating in weak inversion. Applying the same calculations to the characteristic of the conventional CMOS output stage with transistors operating in strong inversion gives

$$i_{AB} = \frac{1}{4}\left(1 + \sqrt{\frac{I_2}{I_1}} \right) i_{11} + \frac{1}{4}\left(1 + \sqrt{\frac{I_1}{I_2}} \right) i_{12} \tag{3-41}$$

where the equation is only valid for I_1, I_2 smaller than $4I_Q$. For the bipolar class-AB characteristic with minimum current control the following transfer function is found

$$i_{AB} = \left(\frac{I_2}{I_1 + I_2}\right)^2 i_{11} + \left(\frac{I_1}{I_1 + I_2}\right)^2 i_{12} \tag{3-42}$$

Comparing the above equations with equation (3-37), expressions for k_1 and k_2 can be found. In the quiescent state, the values of both k_1 and k_2 are $^1/_2$ if the conventional characteristic is used, and $^1/_4$ if the characteristic with minimum current is applied. In order to guarantee stability for all values of the output current, it is important to know how k_1 and k_2 change as a function of the output current. Or, more generally, it is important to understand how the changes of the output current affect the unity-gain frequency as given by equation (3-38). After inspection of equation (3-38), we conclude that we should investigate the products $k_1 g_{m11}$ and $k_2 g_{m12}$. The drain current I_1 of M_1 can be expressed as a function of the quiescent current by

$$I_1 = \alpha I_Q \tag{3-43}$$

By substituting equation (3-43) in the class-AB relation the value of I_2 can be found, while the transconductance of transistors M_1 and M_2 can be expressed as a function of the transconductance in the quiescent state by defining parameters $g_{m11,Q}$ and $g_{m12,Q}$ which represent the transconductance in the quiescent state of M_{11} and M_{12} respecitvely. Applying these calculations to the conventional class-AB characteristic for bipolar as well as CMOS technology in strong inversion yields

$$k_1 g_{m11} = \frac{1}{2} g_{m11,Q} \quad \text{and} \quad k_2 g_{m12} = \frac{1}{2} g_{m12,Q} \tag{3-44}$$

which are independent of α! The unity-gain frequency of the class-AB loop is found as

$$\omega_{0,AB} = \frac{1}{2} g_{AB} r_{AB} \cdot \left(\frac{g_{m11,Q}}{C_{M1A} + C_{GS1}} + \frac{g_{m12,Q}}{C_{M1B} + C_{GS2}}\right) \tag{3-45}$$

which is in first order independent of the output current. This is an important finding since it means that the unity-gain frequency is well controlled even if the output stage is driven hard. Therefore, a stable class-AB loop can be obtained without too many difficulties. For the class-AB control with minimum current we find

$$k_1 g_{m11} = \frac{1}{4\alpha} g_{m11,Q} \text{ and } k_2 g_{m12} = \left(\frac{1}{2} - \frac{1}{4\alpha}\right) g_{m11,Q} \tag{3-46}$$

which depends on the output current. However, the sum of $k_1 g_{m11}$ and $k_2 g_{m12}$ is independent of α. Thus, assuming that $C_{M1A}+C_{GS1}$ is equal to $C_{M1B}+C_{GS2}$, the unity-gain frequency of the class-AB loop of the bipolar circuit with minimum current is

$$\omega_{0,AB} = \frac{1}{2} g_{AB} r_{AB} \cdot \frac{g_{m11,Q}}{C_{M1} + C_{GS}} \tag{3-47}$$

where C_{M1} is the average value of the Miller capacitors C_{M1A} and C_{M1B}, and C_{GS} is the average value of the gate-source capacitances C_{GS1} and C_{GS2}. The above assumption is valid in first order because the Miller capacitors usually have equal values for reasons of distortion. The gate-source capacitances of the output stage are usually not equal but are smaller than the Miller capacitors.

Concluding, the unity-gain frequency of the class-AB loop is in first order independent of the output current for most class-AB control circuits. However, this is not always true. For example, the CMOS simple minimum selector using a transistor in the linear mode as shown in Fig. 3-37 does not have a constant bandwidth when the transistors operate in strong inversion. The unity-gain frequency of the class-AB loop of this circuit is a factor $\sqrt{2}$ larger at high output currents than in the quiescent state.

Now that we have found expressions for the unity-gain frequency of the class-AB loop, the next step is to derive requirements for the unity-gain frequency. First of all, the loop must be stable, having sufficient phase margin. Therefore, the unity-gain frequency must be smaller than the frequency of non-dominant poles in the loop. An important pole is created by the output resistance R_{AB} of the control circuit, and the input capacitance C_{AB}. Therefore the first requirement is

$$\omega_{0,AB} < \frac{1}{2} \frac{1}{R_{AB} C_{AB}} \tag{3-48}$$

where the factor $^1/_2$ is used to obtain a phase margin of approximately 60°. Depending on the implementation of the control block, a number of other non-dominant poles may exist requiring a further reduction of the class-AB-loop unity-gain frequency. A second requirement relates the bandwidth of the class-AB loop to the bandwidth of the main amplifier. Since the class-AB loop must be able to correctly bias the output transistors for all frequencies used by the main amplifier, we conclude

$$\omega_{0,AB} > \omega_0 \qquad (3\text{-}49)$$

where ω_0 is the unity-gain frequency of the main amplifier. To fulfill the requirements, the unity-gain frequency of the class-AB loop can be changed by modifying the transconductance g_{AB} of the class-AB amplifier M_{13}, M_{14}. Further reduction can be obtained by inserting compensation capacitor $C_{C,AB}$ as shown in Fig.3-39.

A final property of interest is the coupling between the gates of the output transistors. As discussed in the previous section, ideally a floating voltage source should be connected between the gates of the output transistors so that the input signal is equally shared by the output transistors in the quiescent state. If one output transistor is driven hard, however, all signal should be redirected to the active output transistor. To investigate this behavior the impedance $z_{g1,g2}$ between the gates of the output transistors can be calculated, yielding

$$z_{g1,g2} = \frac{v_{g1} - v_{g2}}{g_{AB} r_{AB} (k_1 g_{m11} v_{g1} - k_2 g_{m12} v_{g2})} \qquad (3\text{-}50)$$

In the quiescent state k_1 and k_2 are, for example, both equal to $^1/_4$ if a class-AB control with minimum current is used, and, for balancing reasons, the transconductance of transistors M_{11} and M_{12} is equal, yielding

$$z_{g1,g2} = \frac{1}{g_{AB} r_{AB} k_1 g_{m11}} = \frac{4}{g_{AB} r_{AB} g_{m11}} \qquad (3\text{-}51)$$

which confirms the low impedance between the two gates. When one output transistor delivers a large output current, for example M_1, the voltage at the gate is regulated constant so that v_{g2} is very small. Further, k_1 is zero and thus the impedance $z_{g1,g2}$ is very large and the active output transistor can handle the large output currents without being limited by the inactive output transistor. All the input signal is steered through the class-AB amplifier to the active output transistor M_1.

We now move on to complete implementations of feedback class-AB biased rail-to-rail output stages. A bipolar feedback-biased class-AB output stage that can operate on low supply voltages is shown in Fig.3-40. Across R_{12} a voltage is created using emitter measuring transistor Q_{12} and current mirror Q_{22}, Q_{24} that is proportional to the collector

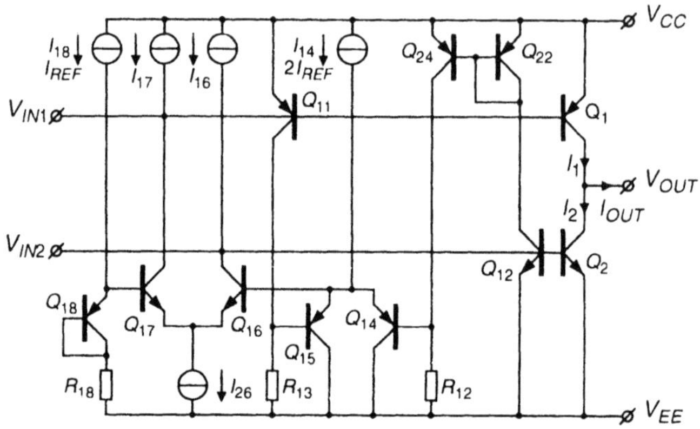

Fig. 3-40 *Low-voltage bipolar feedback-biased class-AB output stage with minimum selector.*

current of Q_2. The voltage across R_{13} is generated using measuring transistor Q_{11} and is proportional to the collector current of Q_1. The output voltage of decision pair Q_{14}, Q_{15} is controlled by the smaller of the two voltages across resistors R_{12} and R_{13}. The smaller of the two voltages corresponds to the smaller of the collector currents of the output transistors. The decision pair drives class-AB amplifier Q_{16}, Q_{17} which completes the feedback loop by driving the bases of the output transistors. This way, a feedback loop is created that controls the minimum current of the output stage. Using this control circuit the quiescent current is given by

$$I_Q = \frac{A_1}{A_{11}} I_{REF} \qquad (3\text{-}52)$$

if the emitter areas of transistors Q_{18}, Q_{14}, and Q_{15} are equal, and the voltages across resistors R_{18}, R_{12}, R_{13} are equal in the quiescent state. The minimum current is found as

$$I_M = \frac{1}{2} \frac{V_{R12} - V_T \ln 2}{V_{R12}} I_Q \qquad (3\text{-}53)$$

where V_T is the thermal voltage and V_{R12} is the voltage across resistors R_{12} and R_{13} in the quiescent state. If V_{R12} is 100 *mV*, I_M is $0.4 I_Q$ at room temperature. The class-AB characteristic is shown in Fig.3-41. The output stage can operate on supply voltages down to 1 V. The same principle can be applied in CMOS as shown in Fig.3-42 [13, 14]. However, this circuit has two drawbacks. First, due to the bad matching between the

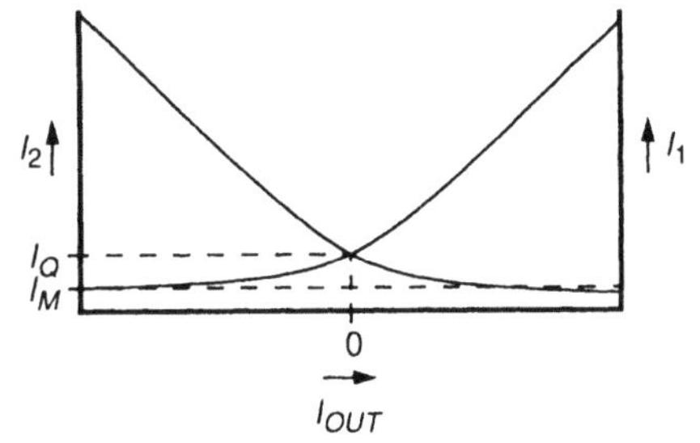

Fig. 3-41 *Class-AB characteristic of low-voltage bipolar feedback-biased output stage.*

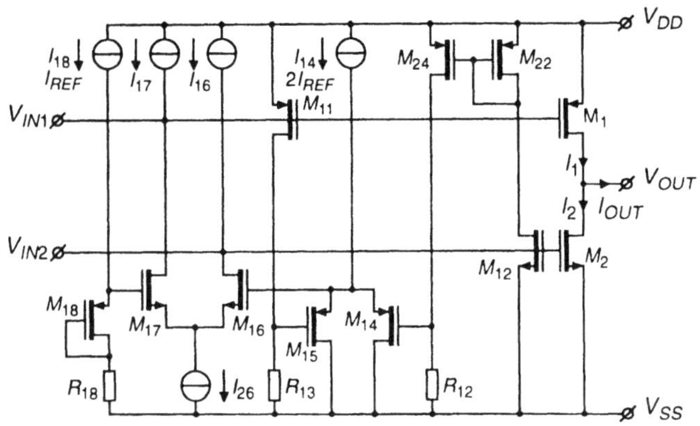

Fig. 3-42 *CMOS low-voltage feedback-biased class-AB output stage with minimum selector.*

gate-source voltage of NMOS and PMOS transistors, the voltage across current source I_{26} is not well determined. To guarantee proper operation of I_{26} sufficient headroom must be created to allow for process variations. Therefore, the minimum supply voltage of the circuit is increased. Secondly, because of the resistors in the translinear loop, it is not possible to fix both the quiescent current and the minimum current accurately. If all the resistors are equal and the gate-source voltages of M_{14}, M_{15} are equal to M_{18} in the quiescent state, the quiescent current is accurately determined

$$I_Q = \frac{W_1}{L_1}\frac{L_{11}}{W_{11}}I_{REF} \qquad (3\text{-}54)$$

where it is assumed that the scaling between M_1 and M_{11} is equal to the scaling between M_2 and M_{12}.

Then, the minimum current is given by

$$I_M = \left(1 - \frac{\sqrt{2}-1}{R_{12,\,13}}\sqrt{\frac{2L_{14,\,15}}{\mu_p C_{ox} W_{14,\,15} I_{REF}}}\right) I_Q \qquad (3\text{-}55)$$

where $R_{12,13}$ is the value of resistors R_{12} and R_{13} and $W_{14,15}/L_{14,15}$ the W/L ratio of transistors M_{14} and M_{15}. Clearly, the minimum value is influenced by the resistors and the gate-source voltage of M_{14} and M_{15}. By making R_{12} and R_{13} large enough, a reasonable amount of minimum current can be secured as demonstrated by the class-AB characteristic shown in Fig.3-43. The first drawback can be solved by replacing PMOS

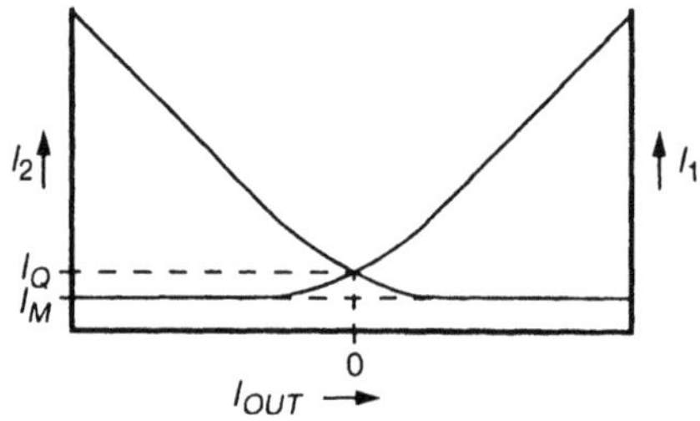

Fig. 3-43 *Class-AB characteristic of CMOS low-voltage feedback-biased class-AB output stage.*

transistors M_{14} and M_{15} by two diode coupled NMOS transistors, as shown in Fig.3-44 [14]. Assuming again that the gate-source voltage of M_{14} and M_{15} is equal to the gate-source voltage of M_{18}, the quiescent current is equal to

$$I_Q = \frac{W_1}{L_1}\frac{L_{11}}{W_{11}}(I_{REF} - I_{REF2}) \qquad (3\text{-}56)$$

and the minimum current

$$I_M = \left(1 - \frac{I_{REF2}}{I_{REF} - I_{REF2}}\,\frac{\sqrt{2}-1}{R_{12,\,13}(I_{REF} - I_{REF2})}\sqrt{\frac{2I_{REF2}L_{14,\,15}}{\mu_n C_{ox} W_{14,\,15}}}\right) I_Q \qquad (3\text{-}57)$$

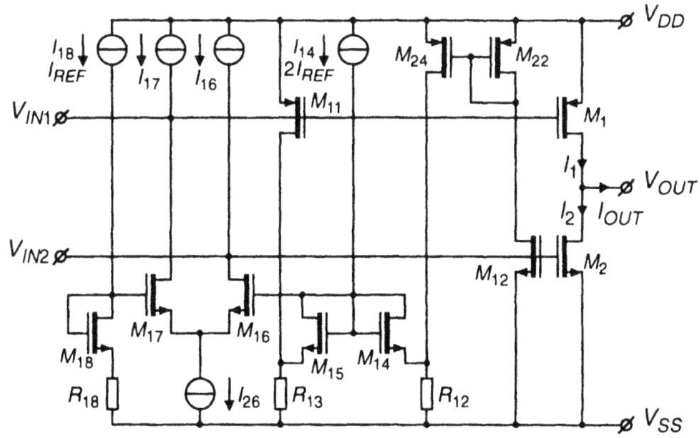

Fig. 3-44 *CMOS low-voltage feedback-biased class-AB output stage with diode-coupled minimum selector.*

Now the reference current I_{14} also flows through resistors R_{12} and R_{13}, counteracting the setting of the minimum current. Thus, the reference current must be small compared to the total current through R_{12} and R_{13} in order to maintain a minimum current. The resulting class-AB characteristic is plotted in Fig. 3-45.

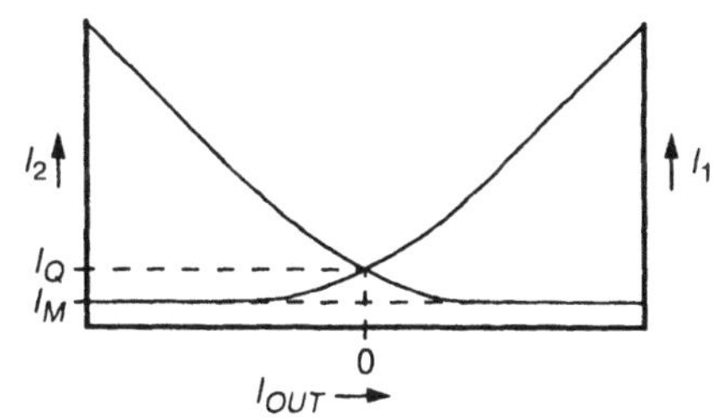

Fig. 3-45 *Class-AB characteristic of CMOS low-voltage feedback-biased class-AB output stage with diode-coupled minimum selector.*

A CMOS feedback circuit without resistors exists [15, 14] but it uses a folded structure consuming more quiescent current than the previous output stage. Recently, a much simpler and more compact minimum selector circuit has been proposed [11] that does not employ resistors. An output stage which applies this minimum selector in a feedback class-AB circuit is shown in Fig. 3-46 [16, 17]. The drain current of M_{12} is proportional to the drain current of M_1. The drain current of M_{12} flows through mirror M_{22}, M_{24} which is part of the minimum selector circuit M_{11}, M_{22}, M_{24} already discussed in this

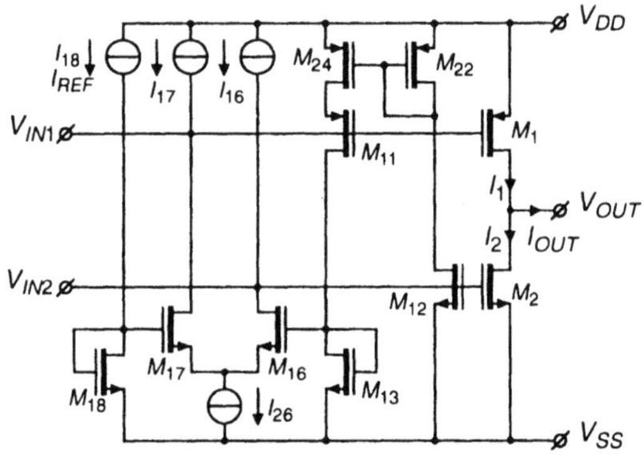

Fig. 3-46 *CMOS leve-voltage feedback-biased class-AB output stage using simple in vnnum selector.*

section and depicted in Fig.3-37. The gate of M_{11} is connected to the gate of output transistor M_1, so that transistor M_{24} operates in the linear mode when the output stage is in the quiescent state. The drain current of M_{11}, which is the output of the minimum selector, flows through M_{13} and steers the class-AB amplifier M_{16}, M_{17}. The class-AB amplifier regulates the signals at the gates of the output transistors in such a way that the current through M_{13} is equal to the reference current I_{REF} that flows through M_{18}. If the output stage is in the quiescent state, the drain current of M_1 is equal to the drain current of M_2. In this situation, transistors M_{11}, M_{24}, and M_{22} should be designed such that the voltage at the gate of M_{11} is equal to the voltage at the gates of M_{24} and M_{22}. Then, transistor M_{24} operates in the linear mode, and the combination M_{11}, M_{24} can be considered as a single transistor with a double length. Therefore, the current through M_{22} and M_{12} is twice as large as the current through M_{13} which is regulated to I_{REF} by the class-AB amplifier. The quiescent current is now given by

$$I_Q = 2\frac{W_2 L_{12}}{L_2 W_{12}} I_{REF} \tag{3-58}$$

When M_1 delivers a large output current, its gate-source voltage will be large and the voltage between the positive rail and the source of M_{11} will be sufficient to allow M_{24} to operate in saturation. Transistors M_{22}, M_{24}, M_{11} now function as a cascoded current mirror and simply mirror the current of measuring transistor M_{12} into M_{13}. Thus, the bias current of the inactive transistor M_{12} is regulated at a constant value equal to half the quiescent current. When M_2 delivers a large output current, a lot of current flows through M_{12} and M_{22}. Transistor M_{24} then pulls the source of M_{11} to the positive supply rail.

Now M_1 and M_{11} operate as a current mirror and the current of M_1, attenuated by a possible scaling factor of mirror M_1, M_{11}, flows through M_{13}. Consequently, the current of inactive transistor M_1 is controlled and is equal to half the quiescent current. The minimum current of the output stage is given by

$$I_M = \frac{W_2 L_{12}}{L_2 W_{12}} I_{REF} = \frac{1}{2} I_Q \qquad (3\text{-}59)$$

where it is assumed that the scaling between M_1, M_{11} and M_2, M_{12} is equal. Another way to investigate the class-AB behavior is by finding the relation between the drain current I_1 of M_1, the drain current I_2 of M_2, and I_{REF}. The complete relation describing this type of minimum selector is calculated in the beginning of this section as given by equation (3-34). The class-AB plot is shown in Fig.3-47. A good class-AB behavior is

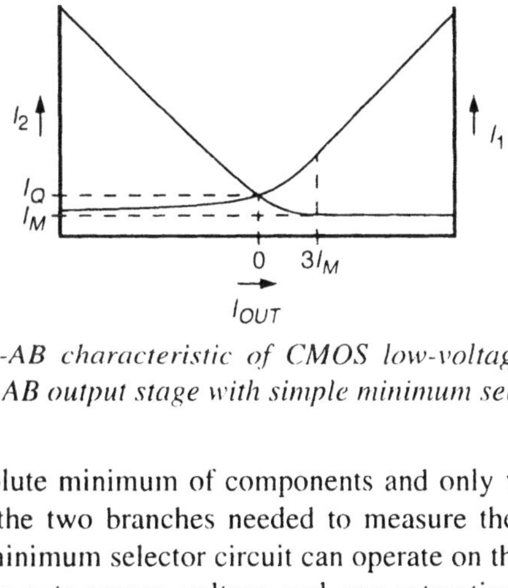

Fig. 3-47 *Class-AB characteristic of CMOS low-voltage feedback-biased class-AB output stage with simple minimum selector.*

obtained using an absolute minimum of components and only very little current, since we only need to bias the two branches needed to measure the currents in the output transistors. Also, this minimum selector circuit can operate on the lowest supply voltage possible since only one gate-source voltage and one saturation voltage is needed. Of course, as used in this example, in combination with a differential control amplifier an additional saturation voltage is required.

An important drawback of the simple minimum class-AB control circuit is that the bandwidth of the class-AB loop changes as a function of the output current, as described earlier in this section. The bandwidth of the class-AB loop when one of the output transistors delivers large output currents is a factor √2 larger than in the quiescent state [17]. In addition, when M_1 handles large output currents, the control loop controlling the minimum current of M_2 contains PMOS mirror M_{22}, M_{24}, and cascode

M_{11} resulting in two poles at a relatively low frequency. Therefore, the phase margin of the class-AB loop reaches a minimum when M_1 drives high output currents.

Another drawback of the circuit shown in Fig. 3-46 is that although it accurately controls the minimum current of the output transistors, the quiescent current depends on how well the factor two is approached by the combination of M_{11} and M_{24}. The factor two may be influenced by several non idealities such as short-channel effects, the back-gate effect, and mismatch between M_{11}, M_{24} and M_{22}. We are, of course, more interested in accurate control of the quiescent current than in accurate control of the minimum current.

Fortunately, we can easily modify the circuit in such a way that the quiescent current is accurately controlled. We can create the bias current of M_{18} using a similar circuit as is applied to measure the bias current of the output transistors. This idea is used in the circuit shown in Fig. 3-48. Using transistors M_{31} and M_{33}, a model of M_{11} and M_{24} is

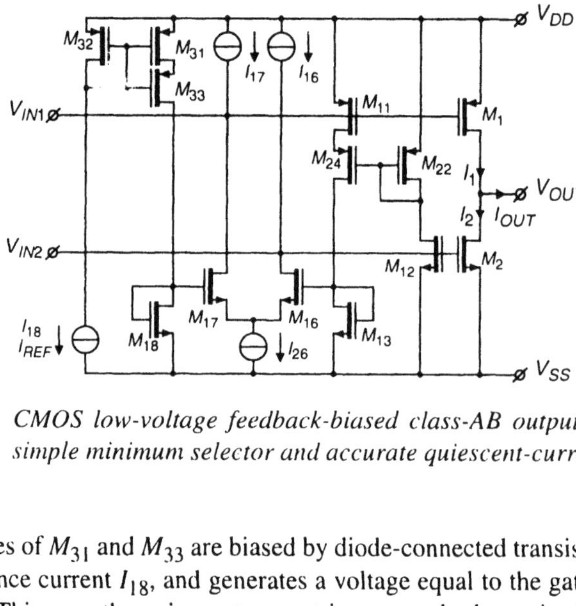

Fig. 3-48 *CMOS low-voltage feedback-biased class-AB output stage with simple minimum selector and accurate quiescent-current control.*

created. The gates of M_{31} and M_{33} are biased by diode-connected transistor M_{32} which is biased by reference current I_{18}, and generates a voltage equal to the gate-source voltage of M_1 and M_{22}. This way, the quiescent current is accurately determined by I_{18}. Another modification shown in Fig. 3-48 is that the function of M_{11} and M_{24} is interchanged: the source of M_{11} is now connected to the positive supply rail while the source of M_{24} is connected to the drain of M_{11} and the output of the minimum selector is now the drain of M_{24}. The advantage of this configuration is that the layout of the output transistors is easier, due to the fact that the source of measuring transistor M_{11} is connected to the source of output transistor M_1. Therefore, we only have to make a separate connection to the drain of M_{11}.

Further, the control loop that controls the minimum current of M_{12} is shorter than the control loop in the circuit shown in Fig.3-46 improving the stability of the class-AB feedback loop. There is now more symmetry in the conrol loop: the control loop of M_{11} contains a cascode while the control loop of M_{12} contains a mirror. If the class-AB loop is carefully designed, the simple minimum selector circuit is one of the most important class-AB feedback control circuits thanks to the high current efficiency and small number of transistors.

It is therefore interesting to consider a similar technique for application in bipolar technology. The bipolar version of the output stage with simple minimum selector is shown in Fig.3-49. All the MOS transistors of the previous circuit are now replaced by

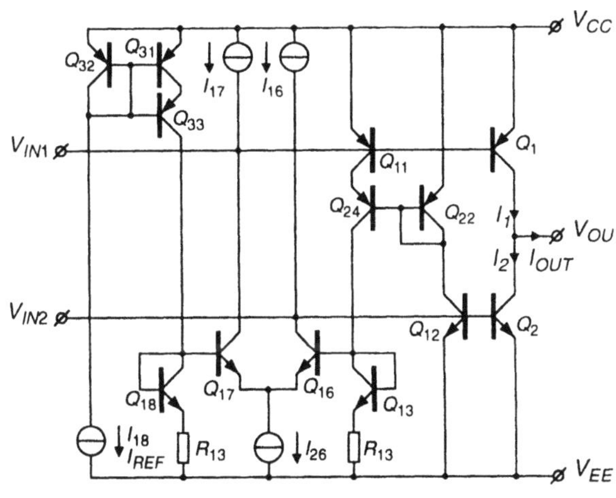

Fig. 3-49 *Simple minimum selector applied in low-voltage bipolar feedback-biased class-AB output stage.*

bipolar transistors. The circuit operates well, since a saturated bipolar transistor shows nearly the same symmetry as found in an MOS transistor operating in the linear mode. This is, for example, expressed by the Ebers-Moll model. Unfortunately, operating a bipolar transistor in the saturation region presents some problems. The base current increases and partly flows into the substrate. The charge carriers flowing into the substrate may cause latch-up and, therefore, careful layout techniques must be applied. The class-AB characteristic is shown in Fig.3-50 and shows remarkable symmetry. This

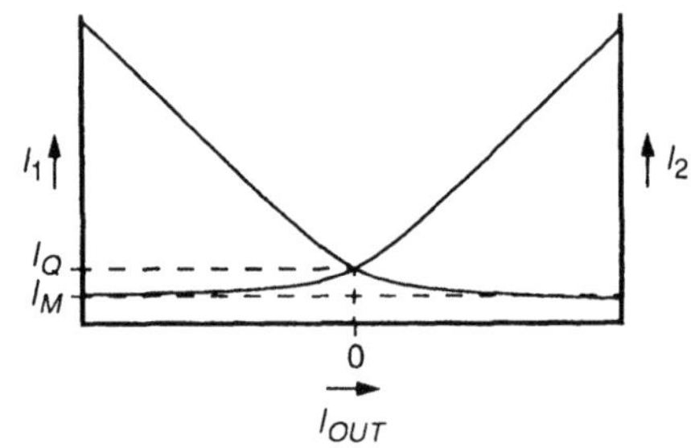

Fig. 3-50 *Class-AB characteristic of bipolar low-voltage feedback-biased class-AB output stage with simple minimum selector.*

is due to the lateral PNP transistors which are, in fact, symmetrical devices. A version of the circuit which applies vertical NPN transistors is depicted in Fig.3-51. To avoid

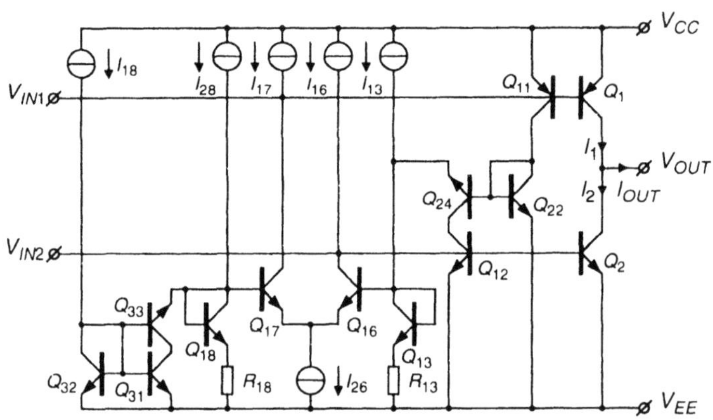

Fig. 3-51 *Bipolar low-voltage feedback-biased class-AB output stage with vertical NPN minimum-selector circuit using transistors operating in the reverse mode.*

influence of the asymmetry of vertical devices, the upper transistor Q_{24} of the minimum selector should operate in the reverse mode, so that its forward biased emitter-collector junction matches with the base-collector junction of Q_{12}. However, because the breakdown voltage of the base-emitter junction is low, of the order of 2V, we must take care that the voltage across the base-emitter junction remains limited. Therefore, a folded structure is used. The output of the minimum selector Q_{12}, Q_{22}, Q_{24}, is connected to foloded NPN diode Q_{13}. A similar structure is used in the reference circuit Q_{31}-Q_{33}, Q_{18}.

Additional current sources I_{13} and I_{28} are required to bias the diodes Q_{13} and Q_{18}. Consequently, some more bias current is needed than in the previous circuits. The class-AB plot is depicted in Fig.3-52 and displays a good symmetry and a high minimum current. Thus, the MOS simple minimum selector can be adapted in order to be used in

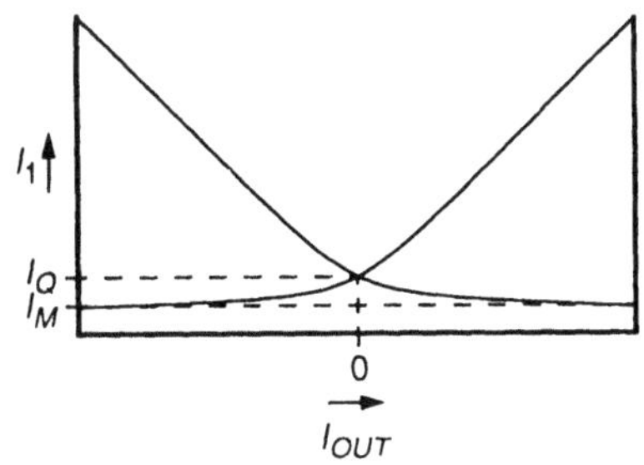

Fig. 3-52 *Class-AB characteristic of bipolar low-voltage feedback-biased class-AB output stage using NPN simple minimum selector with vertical NPN transistors operating in reverse mode.*

bipolar technology. Therefore, compact class-AB control circuits can also be realized in bipolar technology using lateral PNP transistors as well as vertical transistors. However, the circuit must be carefully laid out in order to prevent latch-up problems caused by operating bipolar transistors in the saturation region, or in the reverse mode.

This concludes the discussion of rail-to-rail output stages. In the next section we consider the operation and class-AB biasing of all-NPN output stages.

3.6 All-NPN output stages

About a decade ago, the all-NPN output stage was very popular because of the superior performance of NPN transistors over PNP transistors. Since the arrival of complementary processes complementary stages are often applied. However, due to the additional processing steps, the costs of such processes are high while the NPN transistors still feature better beta, smaller die area and at least a factor three higher bandwidth. Further, due to costs, most bipolar and BiCMOS processes only have lateral PNP transistors available which consume much larger die area and have much smaller bandwidth than the NPN transistors. Therefore, the all-NPN output stage has an important advantage in wideband amplifiers and amplifiers with high output current. Because the output voltage range is one base-emitter voltage smaller than the output voltage range of a rail-to-rail output stage, the all-NPN output stage is not suited to application at very low supply voltages. But at voltages down to 1.8 *V*, the reduced output range can easily be compensated by the reduction in chip area and the much higher bandwidth.

Before discussing the class-AB control of the all-NPN output stage, we first consider the signal behavior because this looks somewhat puzzling at first sight. As can be concluded from the circuit shown in Fig.3-53, where a two-stage amplifier is shown consisting of all-NPN output stage Q_1, Q_2 and input stage Q_3, Q_4, the two output transistors have completely different functions, because the upper one Q_1 is connected as a voltage follower while the lower one Q_2 is connected as an inverting amplifier. To obtain a

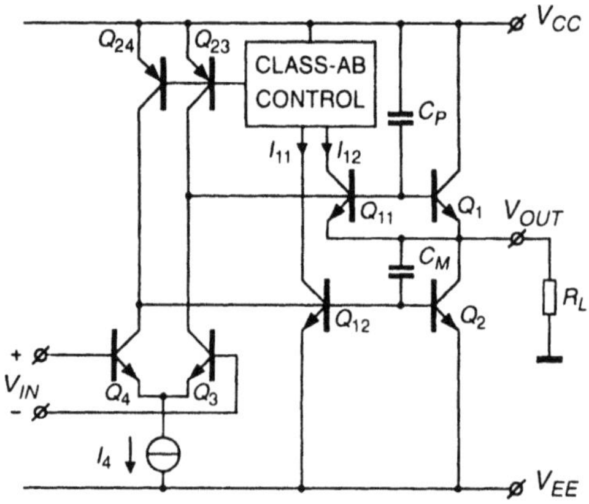

Fig. 3-53 *All-NPN output stage driven by differential input stage showing principle of class-AB control.*

balanced output stage we must somehow ensure that the behavior of both halves is equal. A key step in the balancing is the observation that both output transistors are driven by a current source. If the output impedance of Q_3, Q_4 and Q_{23}, Q_{24} is infinite, the current drive is ideal so that the bases are completely isolated. Therefore, interchanging the emitter and collector connection of the output transistors only results in a signal inversion. Thus, the signal behavior of both output transistors is equal if we make C_P equal to C_M [8]. The same conclusion results after calculating the transfer functions of the voltage follower and the inverter, assuming the transistors are driven by currents. The transfer function of the voltage follower is given by

$$\frac{V_{OUT}}{I_{IN}} = \frac{\beta_F}{1 + j\omega\beta_F C_P} \tag{3-60}$$

assuming the transistor current gain β_F is much larger than unity, and the influence of non-dominant poles can be neglected. The transfer function of the inverter is given by

$$\frac{V_{OUT}}{I_{IN}} = -\frac{\beta_F}{1 + j\omega\beta_F C_M} \tag{3-61}$$

Clearly, the behavior of the output stage can be balanced by making C_P equal to C_M. Of course, parasitics connected between the base of Q_1 and ground have to be taken into account. A more detailed calculations reveals that balanced behavior is possible up to very high frequencies [18], as is discussed in the chapter 5.

The class-AB biasing of the output transistors using a feedforward control circuit is very difficult because Q_1 is completely floating. Therefore, feedback biasing is the only principle that yields practical implementations. The feedback principle of rail-to-rail output stages can be easily modified for applications in all-NPN output stages by noting that the only difference between rail-to-rail output stage and all-NPN output stage is that the upper transistor of the all-NPN output stage does not invert the signal. Therefore, the functions of differential drive and in-phase drive are interchanged compared to a rail-to-rail output stage. As shown in Fig.3-53, the signal is delivered to the output transistors using a differential input stage Q_3, Q_4 while the class-AB control generates two in-phase currents created by Q_{23} and Q_{24}. Transistors Q_{11} and Q_{12} measure the base-emitter voltage of the output transistors Q_1 and Q_2 respectively. The collector currents of Q_{11} and Q_{12} are proportional to the collector currents of the output transistors, and are fed into a class-AB control block that implements the desired class-AB function. The class-AB control block drives transistors Q_{23} and Q_{24} which complete the control loop by controlling the base-currents of the output transistors. Thus, the bias currents of the output stage are controlled in such a way that the class-AB relation is satisfied. An example of a complete implementation is depicted in Fig.3-54. [9,19,18] The control block is implemented using a minimum selector circuit similar to the circuit shown in Fig.3-33. The collector currents I_{11} and I_{12} of transistors Q_{11} and Q_{12} generate voltages across diode-connected transistors Q_{13} and Q_{14}. The voltages across the diode-connected transistors drive minimum selector Q_{15}, Q_{16}. The output of the minimum selector, the common-emitter node of Q_{15}, Q_{16}, is controlled by the higher of the two input voltages at the bases of the minimum selector transistors. Since the higher of the two voltages at

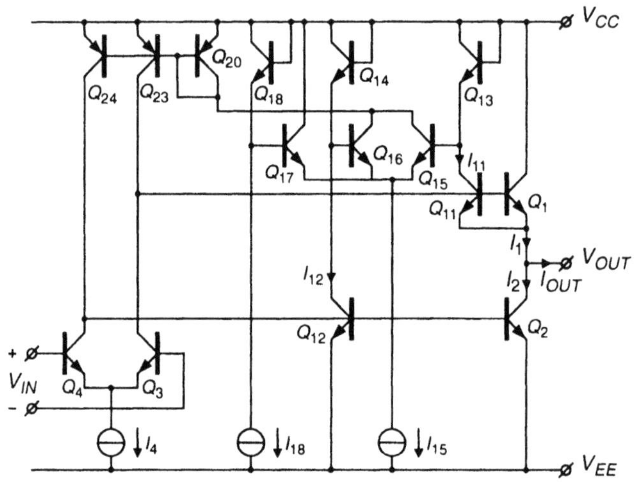

Fig. 3-54 *All-NPN output stage with feedback class-AB control using minimum selector driven by differential input stage.*

the bases corresponds to the lower of the voltage drops across the diodes, the output of the minimum selector is controlled by the smaller of the two collector currents of the output transistors. The minimum selector circuit forms a differential pair together with transistor Q_{17}. The differential pair Q_{15}-Q_{17} compares the minimum of the two voltages across Q_{13} and Q_{14} to a reference voltage created by Q_{18} and I_{18}. The feedback loop is

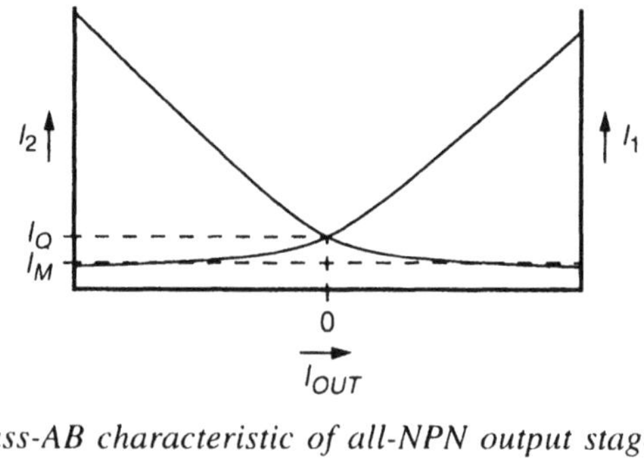

Fig. 3-55 *Class-AB characteristic of all-NPN output stage with minimum-selector class-AB control.*

completed by using the output current of the differential pair to control the input currents of the output transistors. In this example, the collector currents of Q_{15} and Q_{16} are used to generate two in-phase currents using current mirror Q_{20}, Q_{23}, Q_{24}. But the other output, the collector of Q_{17}, can also be used. The two in-phase currents control the base currents of the output transistors. The result is that the lower of the collector currents of both output transistors stays above a certain minimum collector current and thus cut-off of the output transistors is avoided. This is illustrated by the class-AB characteristic plotted in Fig. 3-55.

3.7 BiCMOS output-stage configurations

Up untill now we have only considered pure bipolar or pure CMOS output stages. The reason for this is that from the relatively large number of possible BiCMOS configurations only a few have practical importance. This will become clear after a brief overview of BiCMOS output-stage topologies. It turns out that the most promising configuration consists of a cascade of a CMOS intermediate stage and a bipolar output stage. Since such a circuit is already covered by other parts of this work we do not have to spend much time on this topology. The implementation of bipolar output stages including the class-AB control is extensively discussed in the previous sections, while the frequency compensation techniques required to compensate a cascade of stages are presented in chapter 4. Cascaded BiCMOS topologies are dealt with in more detail in the chapter on realizations, chapter 5. Thus, in this section, we only consider the basic topologies of BiCMOS output stages.

If we limit our attention to rail-to-rail output-stage configurations because of the low-voltage requirements, and only use single transistors or Darlington composite transistors, all possible configurations are presented in Fig.3-56 and Fig.3-57. The first attractive

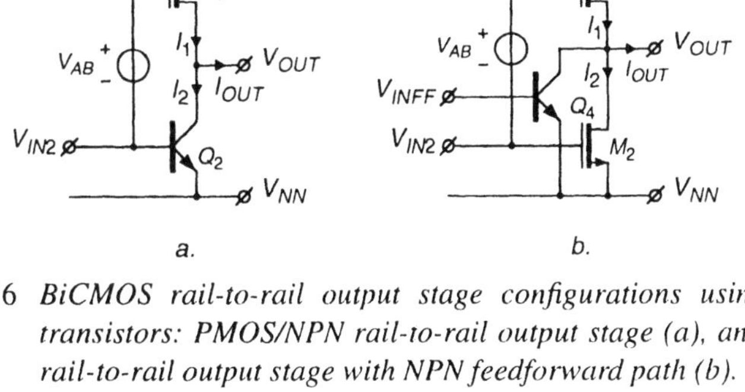

Fig. 3-56 *BiCMOS rail-to-rail output stage configurations using single transistors: PMOS/NPN rail-to-rail output stage (a), and CMOS rail-to-rail output stage with NPN feedforward path (b).*

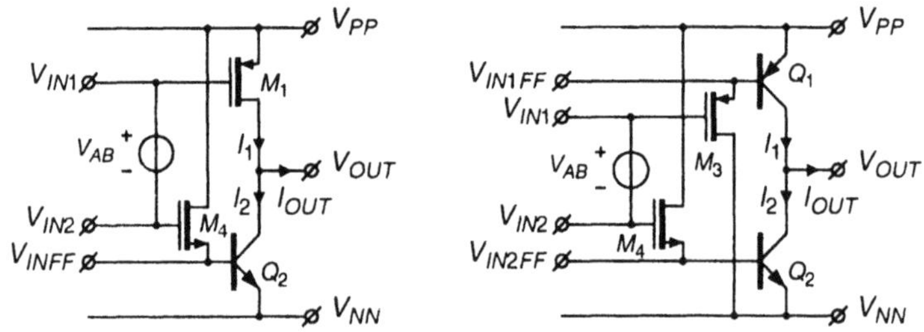

a. b.

Fig. 3-57 *BiCMOS rail-to-rail output stage configurations using Darlington transistors: PMOS/N-type Darlington rail-to-rail output stage (a), and BiCMOS Darlington rail-to-rail output stage (b).*

possibility is to replace the lateral PNP of the bipolar rail-to-rail output stage by a PMOS transistor, yielding the PMOS/NPN rail-to-rail configuration as depicted in Fig.3-56a. However, it is quite difficult to obtain a transconductance using the PMOS transistor that is of the order of the transconductance of the NPN transistor. Therefore the bandwidth is still limited and not determined by the NPN transistor. Another consequence of the smaller transconductance of the PMOS transistor is that the distortion is not so good because of the mismatch between transconductance and current-gain of the two output transistors.

A major problem appears if we consider the operation of the class-AB control. Voltage source V_{AB} is indicated in the figure to remind us that the class-AB control effectively connects the input terminals of the two output transistors in parallel. Thus, the high input impedance of the PMOS transistor is destroyed by the low input impedance of the NPN transistor. It is therefore not possible to obtain sufficient gain using this stage in a two-stage design. Improving the gain of this output stage using single transistors is only possible by applying an ordinary CMOS rail-to-rail output stage. If we still want to take advantage of the high transconductance of the bipolar transistor, we might add an independent feedforward path to the output node using an NPN transistor as shown in Fig.3-56b. The independent feedforward path increases the transconductance of the output stage in the quiescent state and thereby increases the bandwidth. Assuming the output stage is part of a two-stage design, it cannot improve the handling of high output currents, however. Therefore, it is not interesting to consider the possibility of placing a complete bipolar push-pull stage in parallel with the CMOS output stage. A further complication is that it is difficult to bias the NPN transistor. An example is presented in chapter 5, section 5.4.1.

A real improvement of the current gain of the NPN transistor, while still taking advantage of its high output current capability, can be achieved by using a BiCMOS Darlington combination, as depicted in Fig.3-57a. The high transconductance of the NPN can be fully exploited by using a feedforward path that independently drives the NPN transistor yielding a high bandwidth. Drawbacks are that the circuit is not suited to the lowest supply voltages due to the stacking of a base-emitter voltage and a gate-source voltage, and that the distortion is high due to the asymmetry in gain between the upper half of the output stage and the lower half. The last disadvantage can be solved by applying a BiCMOS Darlington in the upper half of the circuit, yielding the BiCMOS Darlington rail-to-rail output stage as shown in Fig.3-57b. This output stage combines high current gain with high output-current capability at moderate supply voltages. Again, the transconductance of the output transistors can be fully exploited by driving the bipolar transistors by independent feedforward paths. The class-AB control recommended to be used with this circuit is a BiCMOS version of the feedforward diode-coupled class-AB control which is shown in Fig.3-20, applied in a bipolar Darlington output stage. An example is discussed in chapter 5, section 5.4.1.

Thus, using single transistors or composite transistors, that nearly behave as a single transistor, like Darlingtons, the only interesting solutions are the BiCMOS Darlington circuit shown in Fig.3-57b, which cannot be used at very low supply voltages and may be the rather awkward circuit depicted in Fig.3-56b. For good solutions at low supply voltages we should consider cascading of stages. An overview of cascaded BiCMOS output stages is given in Fig.3-58. The first option, depicted in Fig.3-58a, is a simple cascade of a CMOS common-source stage and a bipolar rail-to-rail output stage. Using a feedforward path, full profit can be taken of the high transconductance of the bipolar transistors and the high current gain of CMOS transistors. Also, this configuration is suited for application at the lowest supply voltages. However, due to the cascading of two inverting amplifier stages, frequency compensation is difficult. Therefore, a current mirror is usually inserted between the CMOS intermediate stage and the bipolar output stage, as shown in Fig.3-58b, so that the conventional nested-Miller-compensation scheme can be used.

A final circuit to be considered in this chapter is a circuit which applies the all-NPN output stage. Although the reduced output voltage range restricts its use to voltages down to about 2 V, the bandwidth that can be obtained is very high because the lateral PNP with an f_T of the order of 100 MHz is not used. An interesting circuit arises if we drive the all-NPN output stage by an intermediate stage consisting of PMOS transistors. Although frequency compensation is difficult because the lower half M_4, Q_2 consists of two inverting stages, the PMOS intermediate stage gives several advantages. First of all, the PMOS transistors are the only transistors that handle the large base currents of the output stage, since we do not need a current mirror between M_4 and Q_2. Secondly, the

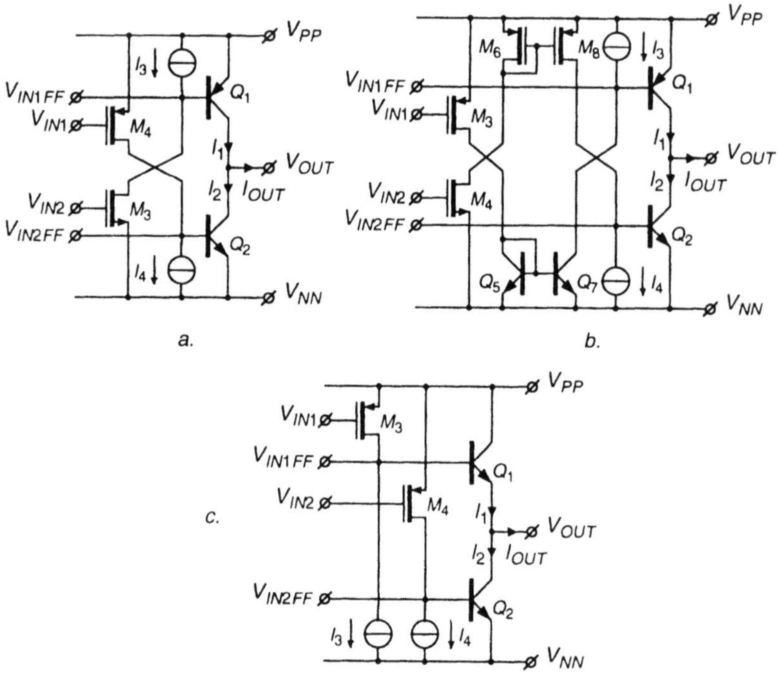

Fig. 3-58 *BiCMOS output stages consisting of a cascade of a CMOS stage and bipolar a bipolar stage: cascade of CMOS stage and bipolar rail-to-rail stage (a),* cascade of *CMOS stage and bipolar rail-to-rail stage with current mirrors (b) and* cascade of PMOS stage and all-NPN stage (c).

PMOS intermediate stage can be simply driven by a differential input stage since the signals at the inputs of the intermediate stage are out of phase. Thus, a summing circuit that sums the differential signals of the input stage is not required. Finally, the two inputs of the PMOS intermediate stage are at the same voltage level so that the Early effects of the driving transistors in the input stage are cancelled. The above advantages make the circuit very suited to application in wideband high-gain designs, or in designs with very high output currents.

3.8 Protection circuits

In the previous sections the bipolar transistors, and especially the NPN transistors, have been promoted as the best devices for application in output stages. However, there are several problems caused by the junction nature of the bipolar devices that require special care. The first problem is saturation - which should not be confused with the saturation region of MOS transistors - which causes additional delay that may result in oscillations, and which causes substrate currents that may even result in latch-up. A second range of problems is related to the operation of the devices at high current densities and high collector-emitter voltages. Under these operating conditions, high current densities or high power dissipation may occur that destroy the devices. To guarantee reliable operation at any load and signal condition, the bipolar output transistors require protection circuits. In section 3.8.1 the saturation of bipolar transistors is discussed and methods to limit saturation are presented. Thereafter, the high-voltage and high-current operation of bipolar transistors is described and circuit solutions to prevent destruction are given in section 3.8.2. The main problem of the limitation circuits is to avoid oscillations in the limiter circuits themselves by keeping the limiter control loops as small as possible.

3.8.1 Saturation protection

Saturation of the ouput transistors should be prevented to avoid oscillations and latch-up. If the collector emitter voltage of a bipolar transistor approaches zero, the base-collector junction becomes forward biased. The base current of the transistor increases to accommodate the reduction of the collector-emitter voltage but eventually the collector-emitter voltage remains constant. Saturation causes the build-up of charge in the base which must be removed before the collector emitter voltage can rise again. The removal of charge causes delay which can result in oscillations. Since most of the base current is injected into the substrate, large voltage drops across the substrate can occur, especially if the substrate is high ohmic. The voltage drop can forward bias other junctions on the chip that are supposed to be reverse biased resulting in latch-up. As discussed in chapter 1 saturation mainly depends on the collector current, the collector resistance, and the temperature. Preferably, the saturation detector should take the variations of these parameters into account.

There are basically three methods that can be used to detect saturation. First, we can directly measure the collector emitter voltage; this is the easiest method. However, we should be cautious because of the voltage drop across the collector-bulk resistance. Because of this, it is not possible to measure the saturation of the internal transistor accurately. Further, we should use a PTAT voltage as reference in order to match the temperature dependence of the saturation voltage. A second method can be based on the measurement of the current gain of the transistor by comparing base current and collector current of the output transistor. However, this is not so easy to implement and, it is difficult to achieve good operation over the range of collector currents. A final method detects the substrate current by measuring voltage drops across the substrate. The functioning of such a circuit depends on the resistivity of the substrate, the position of the substrate contact, and layout of the output transistor and surrounding components. Thus, the first method seems to have the least complications.

A simple way to limit the collector emitter voltage is by connecting a clamp Q_{21} to the collector of output transistor Q_1, as shown in Fig. 3-59 where a Darlington output stage Q_1, Q_2 is used as an example. When the collector voltage reduces below a certain value

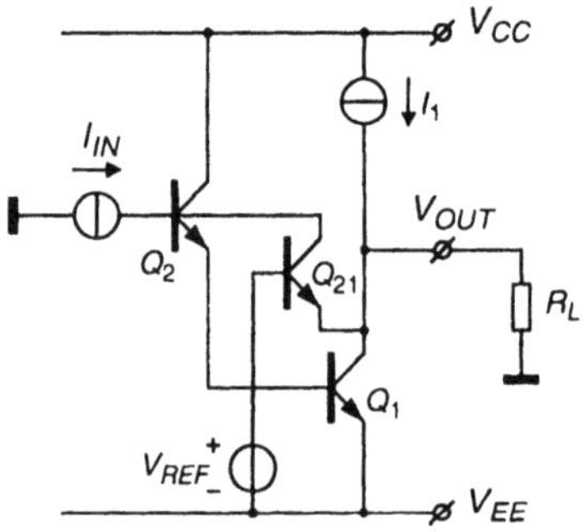

Fig. 3-59 *Saturation protection using NPN clamp applied in Darlington output stage with possible breakdown of base-emitter junction of the clamp.*

determined by the reference voltage, the clamp removes the base current of Q_2. Thus, there is no more current available to drive the output stage into deeper saturation. A major problem of the NPN clamping transistor is that at high collector voltages of Q_1 breakdown of the base-emitter junction of Q_{21} occurs.

Breakdown can be avoided by interchanging the collector and emitter connections of Q_{21} so that it is used in reverse mode during clamping as shown in Fig. 3-60 [4,20,21]. Now the base-emitter voltage of clamping transistor Q_{21} is limited to a save value equal to the base-emitter voltage of Q_2. As shown in the figure, we do not need a separate reference voltage but simply use the base voltage of Q_1 as the reference voltage. Since the base-collector junctions of Q_1 and Q_{21} are connected in parallel, transistor Q_{21} directly measures the forward-bias voltage of the collector-base junction of Q_1. If Q_1 starts to saturate, its collector-base junction becomes forward biased. And, this turns on Q_{21} in the reverse mode so that a current flows from the emitter of Q_{21} to its collector

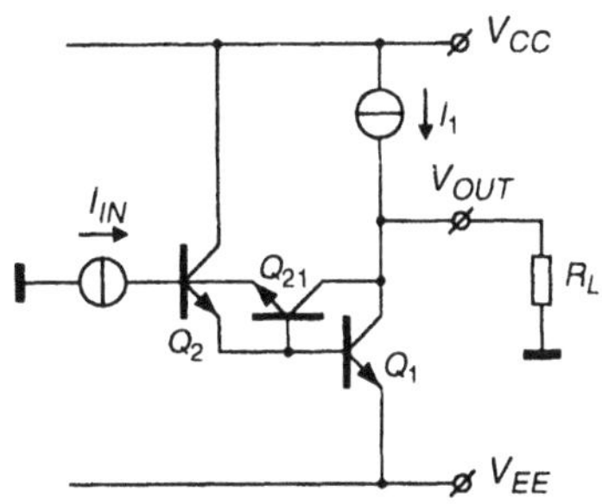

Fig. 3-60 *Saturation protection using reverse NPN transistor applied in Darlington output stage without breakdown.*

removing the base current of Q_2. Thus, deeper saturation of Q_1 is prevented. Due to the good matching between the collector-emitter junction of Q_1 and Q_{21} excellent operation is guaranteed over the temperature range. Also, the control loop is very small assuring sufficient stability.

Disadvantage, however, is that due to the voltage drop across the collector-bulk resistance R_C of Q_1 the saturation detection at higher collector currents is less accurate. Although it is partly compensated by a voltage drop across base resistance the voltage drop across R_C should be taken into account.

3.8.2 Current limitation

The general failure mechanism of transistors operated at high power levels is the temperature rise beyond a certain safe temperature which is of the order of 150 - 200°*C* for silicon. This sets an upper limit for the power dissipation. The maximum operating voltage is limited by avalanche breakdown. Avalanche breakdown occurs in reverse-biased PN junctions when the field strength is so great that the charge carriers gain sufficient energy to free other charge carriers. In bipolar transistors this results in an increase of β. Eventually, β becomes infinite. The avalanche-breakdown mechanism itself is not destructive, but the circuit must be designed such that the current is limited. Two important other mechanisms in bipolar transistors are related to the change of transistor parameters as a function of temperature. If the temperature increases, the parameters change such that the collector current increases. If there is no sufficient cooling the increase of collector current leads to a higher power dissipation which further increases the temperature. Without sufficient cooling the process which is called thermal runaway continues until the device is destroyed. The other thermal effect, known as secondary breakdown, occurs by thermal runaway at a localized point in the transistor. Due to the localized nature, secondary breakdown quickly increases the current density and dissipation until the localized area is destroyed. Further, the resulting temperature gradients may result in a temperature difference between output transistors and fault detection circuits which can result in undetected faults. Therefore, special care must be taken if matching of an output transistor and a transistor in the limiter circuits is required. The above discussed processes lead to a safe operating area as shown in the example in Fig. 3-61 [22,23]. Especially in amplifiers designed for high power dissipation, it is best

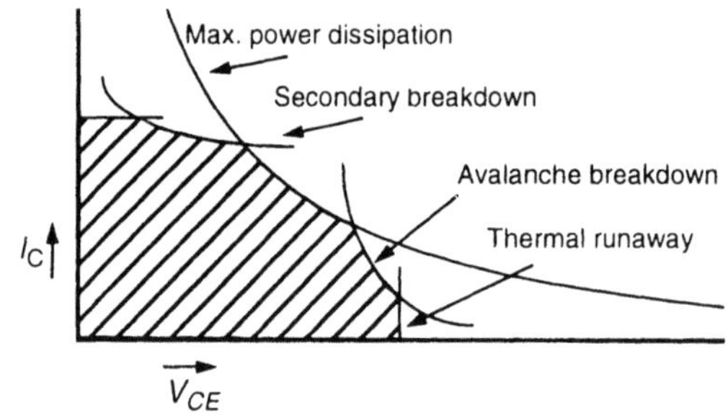

Fig. 3-61 *Safe operating area of bipolar transistors.*

to use the entire safe operating area in order to keep the output transistors as small as possible. Satisfactory circuits based on temperature measurements exist but care must be taken to detect local hot spots [24]. Operational amplifiers, however, are usually designed for low or medium power applications. In that case, it is not necessary to use the complete safe operating area and simple methods to limit the current and the voltage are sufficient. The maximum voltage is limited by using a crow bar to limit the supply voltage and using ESD diodes to limit the voltage at all external pins to the supply voltages. The implementation of a crow bar and ESD circuits is beyond the scope of this work. The maximum current can be limited by special circuits that measure the output current and remove the driving current of the output stage. Such circuits are further discussed in this section. An example of a straightforward current limitation circuit applied in a Darlington output stage is depicted in Fig. 3-62 [4,20]. The collector current of Q_{41} is proportional to the collector current of output transistor Q_1 and generates a voltage drop across R_{42}. As soon as the voltage drop is of the order of a base-emitter voltage, Q_{42} is turned on and the driving current of Q_2 is removed via current mirror Q_{43}, Q_{44}. Thus, a transistor is used to set a threshold. Below the threshold, the base-emitter voltage of Q_{42}, the influence of the limitation circuit is negligible. Above the threshold the influence of the limiter quickly increases. However, because Q_{42} inverts the signal, an additional current mirror Q_{43}, Q_{44} is required. Thus, the control loop is long and it is difficult to prevent oscillations. To shorten the control loop a different arrangement has to be found. An example is shown in Fig. 3-63 [4,20] where the saturation limiter discussed in section 3.8.1 is applied to Q_{41}. Again, a voltage drop is generated across

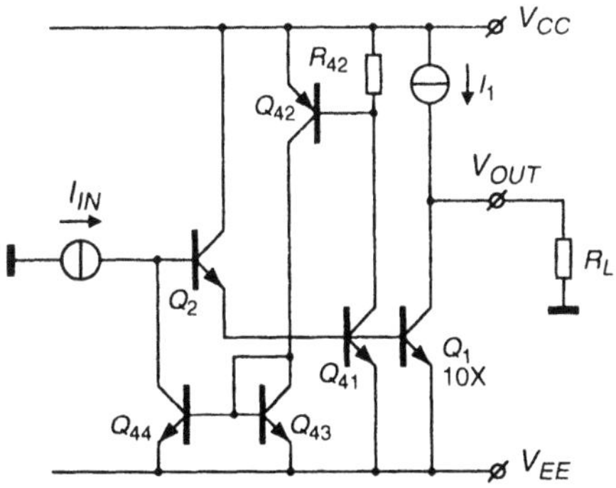

Fig. 3-62 *Straightforward current-limitation circuit using transistor threshold applied in Darlington output stage resulting in a long control loop.*

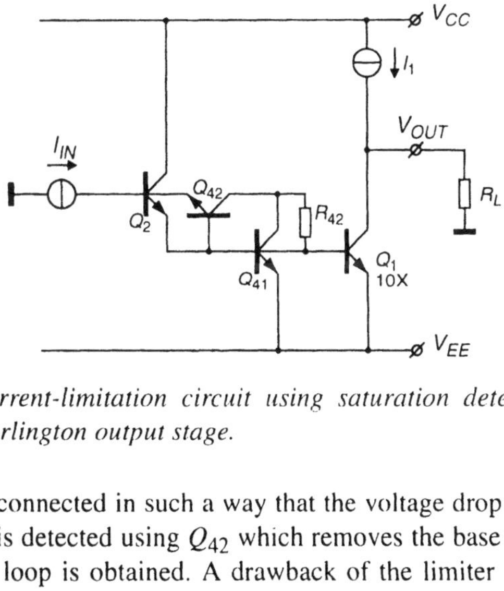

Fig. 3-63 *Current-limitation circuit using saturation detector applied in Darlington output stage.*

R_{42}. Resistor R_{42} is connected in such a way that the voltage drop induces saturation of Q_{41}. The saturation is detected using Q_{42} which removes the base current of Q_2. In this way a short control loop is obtained. A drawback of the limiter is that the low β_R of modern processes seriously limits the strength of the circuit. Therefore another way has to be found to set a threshold without applying a transistor in reverse mode. Another non-inverting device that can set a threshold is a diode. As shown in Fig. 3-64, we can compare the collector current of Q_{41} to a reference current I_{REF} and use diode D_{40} to isolate the limiter from the base of Q_2 if the reference current is larger than the collector current of Q_{41} [20]. In that case part of the reference current flows into diodes D_{48}, D_{49}

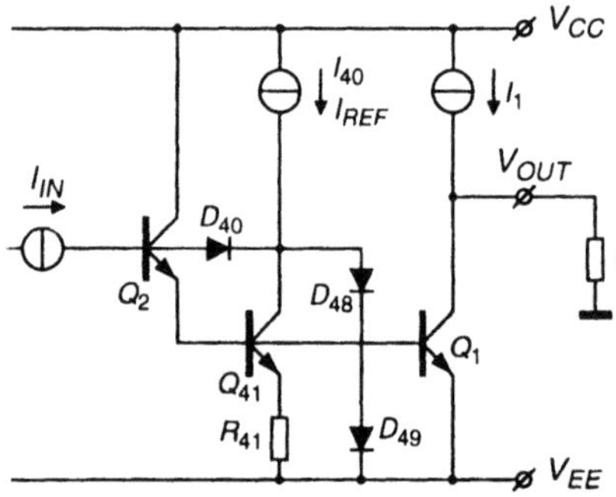

Fig. 3-64 *Current-limitation circuit with short control loop using diode-threshold circuit applied in Darlington output stage.*

to make sure that the voltage at the collector of Q_{41} is correctly biased. As soon as the collector current of Q_{41} becomes larger than I_{REF}, a current flows through diode D_{40} removing the base current of Q_2. This results in a limitation of the output current. In order to reduce I_{REF} to acceptable values, resistor R_{41} has to be inserted which again weakens the limiter. The strength of the limiter can be restored by replacing the resistor with a voltage source or reducing the impedance at the emitter of Q_{41}. As shown in Fig. 3-65, the voltage at the emitter of Q_{41} can be fixed using Q_{42} which is biased by Q_{43}, R_{43} and current source I_{43} [20]. Another improvement is the addition of Q_{45} which limits the

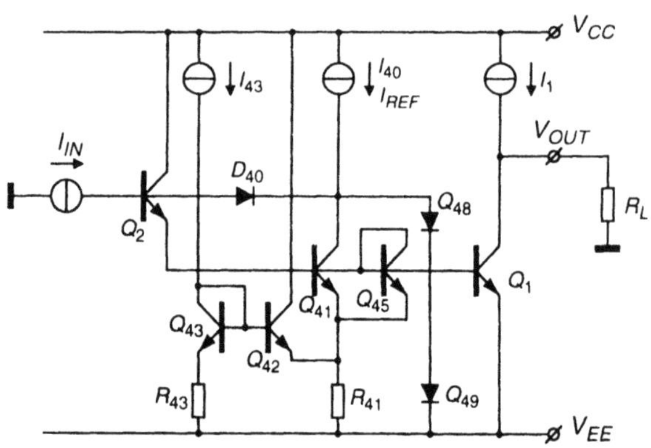

Fig. 3-65 *Improved current-limitation using diode-threshold circuit by fixing the voltage across at the emitter of Q_{41}.*

base-current of Q_1 when the high output current is not caused by a large input current but by breakdown of the output transistor. The limiter can be further improved by completely fixing the voltage at the emitter of Q_{41} as shown in Fig. 3-66, where resistor R_{41} is replaced by Q_{44}. Using the small feedback loop, consisting of Q_{44} and Q_{42}, the

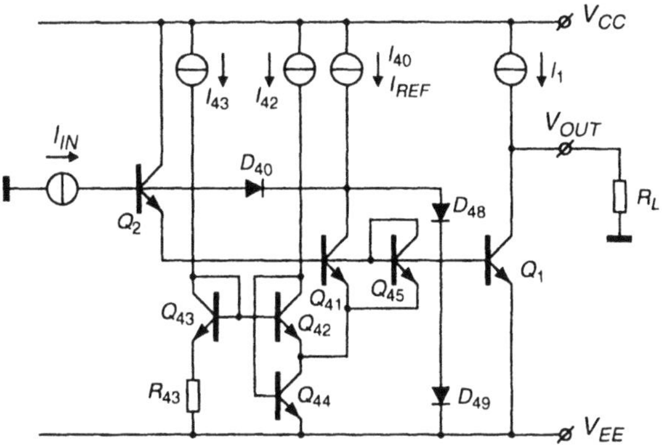

Fig. 3-66 *Further improvement of current-limitation circuit by fixing the voltage at the emitter of Q_{41} using Q_{46}.*

current through Q_{42} is fixed and therefore the voltage at the emitter of Q_{41} is constant. The limitation is demonstrated by the plot of the output current as a function of the input current of the Darlington input stage while applying several versions of the diode-threshold current-limiter circuits. As we see, the simplest version already displays reasonable limitation. Depending on the requirements, such as maximum bias current and maximum input-driving current, one of the improved versions may be needed to limit the output stage within an acceptable range. A final point of concern is the

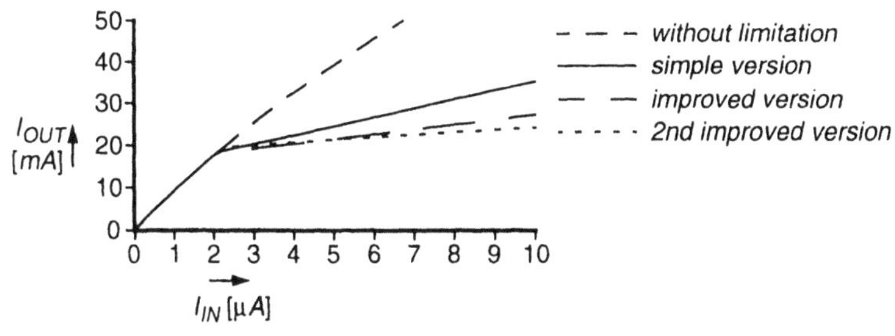

Fig. 3-67 *Plot of limitation of Darlington output stage using diode-threshold parallel current limiter*

temperature behavior. The limitation level of the previous circuits is set using the voltage across R_{43} which equals the voltage at the emitter of Q_{41} at the beginning of the limitation. Thus, we find

$$I_{43} \cdot R_{43} = \frac{kT}{q} \ln\left(\frac{I_{OUT}}{I_{REF}} \cdot \frac{A_{41}}{A_1}\right) \quad (3\text{-}62)$$

and we conclude that I_{43} must be PTAT.

A disadvantage of the previous circuits is that matching of Q_{41} and Q_1 is required. Further, as discussed above, large output currents in localized hot spots may not be detected.

Therefore, in output stages with large dissipation it is better to place a resistor in series with the emitter of Q_1 to correctly measure the output current under all conditions as shown in Fig. 3-66 [25,20]. An additional advantage is that transistors Q_{41} and Q_{42} can

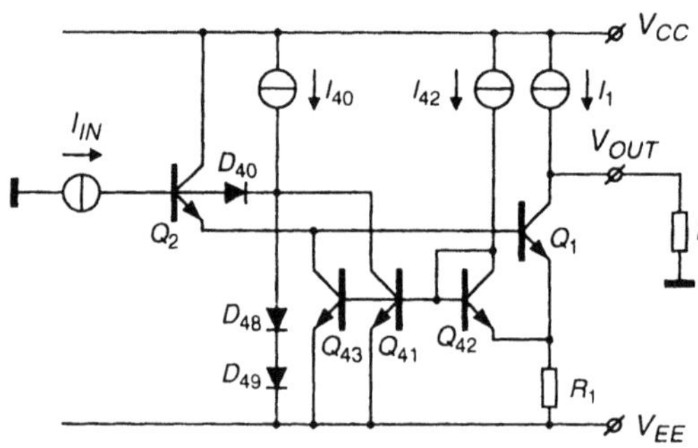

Fig. 3-68 *Current-limitation circuit with short control loop by using resistor in series with the emitter connection of the output transistor.*

have a small emitter area because they do not have to match with Q_1. In low voltage circuits the voltage drop across the R_1 must be small to maintain a rail-to-rail output range. Practical values of the voltage across R_1 are of the order of 18 mV. The voltage across R_1 is transferred to the base of Q_{41} using levelshift diode Q_{42}. If the voltage across R_1 is too high, the collector current of Q_{41} is larger than reference current I_{40}. Then, a current flows through D_{40} removing the base-current of Q_2. Transistor Q_{43} limits the base current of Q_1 when the high output current is caused by breakdown. The operation of the limiter is illustrated by the plot shown in Fig. 3-69. Again, the temperature behavior is important. At the start of limitation we find

$$I_{OUT} \cdot R_l = \frac{kT}{q} \ln\left(\frac{I_{40}}{I_{42}} \cdot \frac{A_{42}}{A_{41}}\right) \tag{3-63}$$

and we conclude that resistor R_l should have a positive temperature coefficient. Therefore, R_l is preferably implemented using metal or low-ohmic poly silicon.

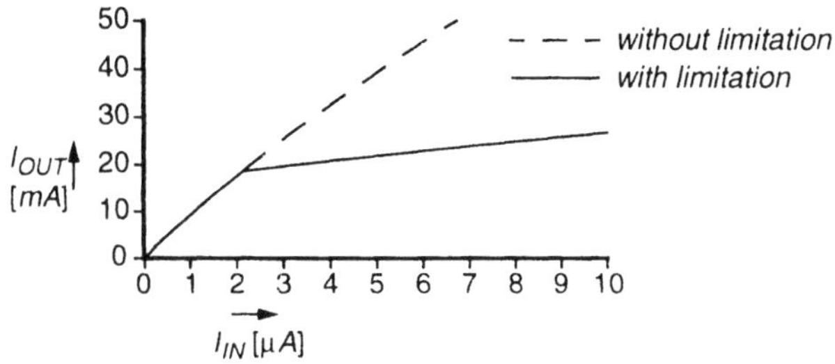

Fig. 3-69 *Plot of limitation of Darlington output stage using series current limiter.*

A final possibility for current limitation is to combine the limitation circuit with the class-AB control circuit as shown in the all-NPN output stage in Fig. 3-70. The collector currents of Q_{11} and Q_{12} are proportional to the collector currents of Q_1 and Q_2, respectively. The collector currents of Q_{11} and Q_{12} flow through transistors Q_{13} and Q_{14} which are connected as diodes if we disregard the resistors R_{13} and R_{14}. Transistors Q_{13} and Q_{14} are part of the minimum-selector class-AB control. The currents also flow through resistor R_{14}. If one of the currents is too large, the voltage drop across resistor R_{14} causes the transistors Q_{13} and Q_{14} to saturate. Due to the saturation the base current increases, generating a voltage drop across R_{13}. The voltage across R_{13} reduces the voltage at the base connections of Q_{15} and Q_{16} so the collector currents of Q_{15} and Q_{16} decrease. Consequently, the current flowing through the current mirror Q_{20}, Q_{23}, Q_{24} decreases, limiting the base current of the output transistors. Since the voltage across R_{14} matches with the collector-base junction, while the voltage across R_{13} matches with the difference of two base-emitter junctions, the temperature coefficient of R_{14} should be negative and the temperature coefficient of R_{13} should be positive. Thus, R_{14} should be implemented using high-ohmic material, while R_{13} should be implemented using low-ohmic material. The advantage of this type of current mirror is that it uses the existing class-AB control circuit and only requires the addition of two resistors. Also, the circuit

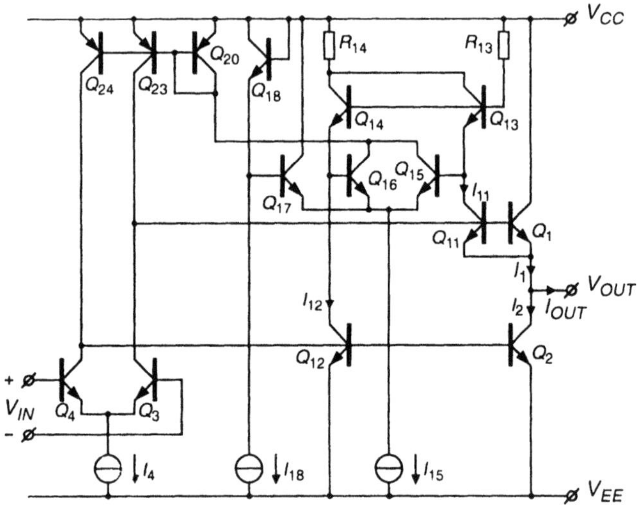

Fig. 3-70 *Current-limitation circuit combined with class-AB control circuit applied in all-NPN output stage.*

protects both output transistors and does not require a separate circuit for each output transistor. The result is a compact and power-efficient limitation circuit.

3.9 Conclusions

In the previous sections a huge number of output stages has been discussed. To give some insight in the application of each circuit this chapter is concluded with a brief overview and a comparison of performance. The overview is divided into two parts. First the basic topologies are considered. Thereafter, the different types of class-AB control circuits are compared. In this chapter we have seen that, due to the low-voltage requirements, rail-to-rail output stages are the most important class of circuits. Table 3-2 gives an overview of the performance of bipolar, CMOS, and BiCMOS rail-to-rail topologies in addition to the performance of the all-NPN topology. Although the all-NPN topology has a smaller output voltage range, it is still favorable due to the high bandwidth and high output-current capabilities. The bipolar rail-to-rail output stage yields a high transconductance, but the performance is limited by the PNP transistor. Thus, it obtains reasonable bandwidth and moderate output current capability. The gain is low which restricts its use to multi-stage designs requiring sophisticated frequency compensation techniques. The CMOS stage obtains moderate bandwidth and output

Table 3-2 *Comparison of basic output stage topologies*

Topology	bipolar rail-to-rail	CMOS rail-to-rail	BiCMOS Darlington rail-to-rail	All-NPN
Gain	-	++	++	-
Bandwidth	+	o	+	++
Output-current capability	o	-	+	++
Output-voltage range	++	++	++	o
Minimum supply voltage	++	++	-	o

++ = excellent, + = good, o = average, - = poor, -- = very poor

current capability due to its lower transconductance. However, because of its high input impedance high gain can be obtained which enables application in two-stage designs. A compromise is found in the BiCMOS Darlington output stage which obtains reasonable bandwidth and output-current capability at a higher supply voltage.

An overview of several performance parameters of bipolar class-AB control techniques is presented in Table3-3. The following bipolar circuits are compared:

1. *Feedforward-biased resistive-coupled output stage (RC,* Fig.3-11*).*
2. *Feedforward-biased separate-resistive-coupled output stage (SRC,* Fig.3-14*)*
3. *Feedforward-biased resistive-coupled output stage with mesh (RCM,* Fig.3-16*)*
4. *Feedforward-biased diode-coupled output stage (DC,* Fig.3-18*)*
5. *Feedforward-biased diode-coupled Darlington output stage (DCD,* Fig.3-20*)*
6. *Feedforward-biased transistor-coupled output stage (TC,* Fig.3-22*)*
7. *Low-voltage feedback-biased output stage (LV,* Fig.3-40*)*
8. *Feedback biased output stage with simple minimum selector (LVS,* Fig.3-49*)*

The current efficiency parameter is mainly effected by folding actions to allow lower supply voltages. The die area relates to the use of large resistors which are applied in the simple feedforward circuits, while resistors of somewhat smaller size are used in the feedback biased circuits. The design complexity factor is mainly influenced by the design effort to stabilize the feedback loop of the feedback biased circuits, the careful layout of substrate contacts surrounding saturating transistors, and the difficult implementation of the required levelshift in the diode-coupled feedforward circuit.

Table 3-3 *Comparison of bipolar class-AB control techniques*

Topology	Feedforward biased						Feedback biased	
	RC	SRC	RCM	DC	DCD	TC	LV	LVS
Gain	+	-	o	o	+	+	+	+
Min. current	-	-	-	o	+	+	+	+
Current consumption	-	+	+	+	+	+	o	++
Min. supply voltage	o	+	+	+	-	-	+	+
Die area	-	-	-	+	+	+	o	+
Design complexity	+	+	o	-	o	+	-	--

++ = excellent, + = good, o = average, - = poor, -- = very poor

Overall, the feedback biased circuits are favorable at very low supply voltages, while the transistor-coupled circuit and the Darlington version of the diode-coupled circuit are interesting options at higher supply voltages. The other feedforward circuits have several disadvantages related to the use of large resistors, but they can operate at low supply voltages and are easy to design.

The following CMOS circuits are compared:

1. *Feedforward-biased resistive-coupled output stage (RC,* Fig.3-11*).*
2. *Feedforward-biased transistor-coupled output stage (TC,* Fig.3-24*)*
3. *Low-voltage feedback biased output stage (LV,* Fig.3-42*)*
4. *Low-voltage feedback-biased output stage with diode-coupled minimum selector (LVD,* Fig.3-44*)*
5. *Low-voltage feedback-biased output stage with simple minimum selector (LVS,* Fig.3-47*)*

The behavior of CMOS class-AB control circuits is summarized in Table3-4. Again, we see that accurate class-AB control with minimum current at low supply voltages can only be obtained using feedback biased circuits, while the feedforward circuits require higher supply voltages but are easier to design.

Table 3-4 *Comparison of CMOS class-AB control techniques*

Topology	Feedforward biased		Feedback biased		
	RC	TC	LV	LVD	LVS
Gain	+	+	+	+	+
Min. current	-	+	+	+	+
Current consumption	-	+	+	o	++
Min. supply voltage	o	-	-	+	++
Die area	-	+	+	o	++
Design complexity	+	+	-	-	-

++ = excellent, + = good, o = average, - = poor, -- = very poor

The final output stage is the all-NPN output stage of which the general properties were already discussed. Since the class-AB control of the all-NPN output stage is only possible using the feedback biased principle there is only one way to fix the class-AB control.

Concluding, in this chapter we have seen that there are a lot of ways to obtain a class-AB output stage in bipolar, CMOS, and BiCMOS topology. Each technique has its own application area and there is no absolute winner among the different circuits. The right technique for each application can be selected with the help of the tables in this section. In general, a CMOS rail-to-rail output stage is used to obtain compact amplifiers where the gain can be obtained using only a few stages, while the bipolar output stage is applied in application with higher bandwidth and higher output currents. The all-NPN output stage is especially suited to extreme situations, very high bandwidth or very high output current. With respect to the class-AB control technique, the feedforward biased transistor-coupled technique shows good behavior at higher supply voltages, while at lower supply voltages feedback biasing is required. In particular, good performance is obtained using the CMOS simple-minimum-selector circuit.

3.10 References

[1] B. Gilbert, "Translinear circuits: a proposed classification", *Electron. Lett.*, Vol. 11, pp. 14-16, Jan. 1975; also "Errata", *ibid.*, p. 136.

[2] W.C.M. Renirie, J.H. Huijsing, "Simplified Class-AB Control Circuits for Bipolar Rail-to-Rail Output Stages of Operational Amplifiers", in *Proc. ESSCIRC*, pp. 183-186, Sept. 21-23, 1992.

[3] W.C.M. Renirie, K.J. de Langen, J.H. Huijsing, "Parallel Feedforward Class-AB Control Circuits for Low-Voltage Bipolar Rail-to-Rail Output Stages of Operational Amplifiers", *Analog Integrated Circuits and Signal Processing*, Vol. 8, pp. 37-48, 1995.

[4] M.J. Fonderie, J.H. Huijsing, *Design of Low-voltage Bipolar Operational Amplifiers*. Boston: Kluwer Academic Publishers, 1993.

[5] D.M. Monticelli, "A quad CMOS single-supply Opamp with rail-to-rail output swing", *IEEE J. of Solid-State Circuits*, Vol. SC-21, pp. 1026-1034, Dec. 1986.

[6] R. Hogervorst, J.H. Huijsing, *Design of Low-voltage Low-Power Operational Amplifier Cells*. Dordrecht, The Netherlands: Kluwer Academic Publishers, 1996.

[7] J.H. Huijsing and D. Linebarger, "Low-Voltage Operational Amplifier with Rail-to-Rail Input and Output Ranges", *IEEE J. of Solid-State Circuits*, Vol SC-20, No. 6, pp. 1144-1150, Dec. 1985.

[8] J.H. Huijsing and F. Tol, "Monolithic Operational Amplifier Design with improved HF behavior", *IEEE J. Solid-State Circuits*, Vol. SC-11, No. 2, pp. 323-328, April 1976.

[9] E. Seevinck, W. de Jager, P. Buitendijk, "A Low-Distortion Output Stage with improved stability for monolithic power amplifiers", *IEEE J. Solid-State Circuits*, Vol. SC-23, pp. 794-801, June 1988.

[10] R.F. Wassenaar, *Analysis of Analog C-MOS Circuits*, Ph.D. thesis University of Twente, The Netherlands, 1996.

[11] E. Seevinck, "CMOS Translinear circuits", in *Advances in Analog Circuit Design: MOST RF Circuits, Sigma-Delta Converters and Translinear Circuits*, W.M.C. Sansen, J.H. Huijsing, R.J. van de Plassche, Eds. Dordrecht, The Netherlands: Kluwer Academic Publishers, 1996, pp. 323-336.

[12] R.F. Wassenaar, private communication.

[13] R. Hogervorst, R.J. Wiegerink, P.A.L. de Jong, J. Fonderie, R.F. Wassenaar, J.H. Huijsing, "CMOS Low-Voltage Operational Amplifiers with constant-gm Rail-to-Rail input stage", in *Proc. IEEE ISCAS*, pp. 2876-2879, May 10-13, 1992.

[14] K.J. de Langen, R. Hogervorst, J.H. Huijsing, "Translinear circuits in Low-Voltage Operational Amplifiers", in *Advances in Analog Circuit Design: MOST RF Circuits, Sigma-Delta Converters and Translinear Circuits*, W.M.C. Sansen, J.H. Huijsing, R.J. van de Plassche, Eds. Dordrecht, The Netherlands: Kluwer Academic Publishers, 1996, pp. 357-38.

[15] R.G.H. Eschauzier, R. Hogervorst, J.H. Huijsing, "A Programmable 1.5 V CMOS Class-AB Operational Amplifier with Hybrid Nested Miller Compensation for 120 dB Gain and 6 MHz UGF", *IEEE J. Solid-State Circuits*, vol. SC-29, no. 12, pp. 1497-1504, Dec. 1994.

[16] K.J. de Langen, J.H. Huijsing, "Compact 1.8V Low-Power CMOS Operational Amplifier Cells for VLSI", in *ISSCC Dig. Tech. Papers*, pp. 346-374, Feb. 1997.

[17] K.J. de Langen, J.H. Huijsing, "Compact Low-Voltage Power-Efficient Operational Amplifier Cells for VLSI", *IEEE J. Solid-State Circuits*, vol. SC-33, no. 10, pp. 1482-1496, Oct. 1997.

[18] K.J. de Langen, R.G.H. Eschauzier, and J.H. Huijsing, "A 1-GHz Class-AB Operational Amplifier with Multipath Nested Miller Compensation for 76-dB Gain", *IEEE J. Solid-State Circuits*, vol. SC-32, no. 4, pp. 488-498, April 1997.

[19] R.G.H. Eschauzier, L.P.T. Kerklaan, and J.H. Huijsing, "A 100-MHz 100-dB Operational Amplifier with Multipath Nested Miller Compensation Structure", *IEEE J. Solid-State Circuits*, vol. SC-27, no. 12, pp. 1709-1717, Dec. 1992.

[20] K.J. de Langen, J. Fonderie, J.H. Huijsing, "Limiting Circuits for Rail-to-Rail Output Stages of Low-Voltage Bipolar Operational Amplifiers", in *Proc. ISCAS*, pp. 1728-1731, April 30 - May 3, 1995.

[21] T. Szepesi, "Design and Circuit Techniques of Integrated Switching Voltage Regulators", in *Advances in Analog Circuit Design: Low-Power Low-Voltage, Integrated Filters and Smart Power*, R.J. van de Plassche, W.M.C. Sansen, J.H. Huijsing, Eds. Dordrecht, The Netherlands: Kluwer Academic Publishers, 1996, pp. 265-291.

[22] E.M. Cherry, D.E. Hooper, *Amplifier Devices and Low-pass Amplifier Design*. New York: Wiley, 1968.

[23] S.K. Ghandi, *Semiconductor Power Devices*. New York: Wiley, 1977.

[24] M. Yamatake, R Widlar, "Dynamic Safe-Area Protection for Power Transistors Employs Peak-Temperature Limiting", *IEEE J. Solid-State Circuits*, vol. SC-22, no. 1, pp. 77-84, Feb. 1987.

[25] H. Zitta, "Smart Power Circuits for Power Switches Including Diagnostic Functions", in *Advances in Analog Circuit Design: Low-Power Low-Voltage, Integrated Filters and Smart Power*, R.J. van de Plassche, W.M.C. Sansen, J.H. Huijsing, Eds. Dordrecht, The Netherlands: Kluwer Academic Publishers, 1996, pp. 249-264.

Overall-Design Techniques

4

4.1 Introduction

In the previous chapters we have seen how to design input stages and output stages. The next step in the design of operational amplifiers consists of determining the required number of stages; for that we consider the main overall parameters of amplifiers. The main overall parameters of operational amplifiers are dc-gain and speed. The gain sets an upper limit for the low-frequencies accuracy of the feedback system in which the amplifier is applied, while speed, in terms of bandwidth, determines the accuracy at higher frequencies. Generally, increasing the number of stages improves the gain but reduces the bandwidth.

Operational amplifiers are usually applied as part of a feedback system. The well-known consequence of applying feedback is the possibility of oscillations. Frequency compensation techniques provide the means to shape the frequency response of the amplifier in such a way that stability can be guaranteed under a variety of feedback factors, load conditions, and parameter changes. The frequency compensation techniques, which are necessary to compensate the amplifier realizations of chapter 5, are treated in this chapter. Next, dynamic distortion, which largely depends on the frequency compensation technique, is analyzed here.

Finally, we establish the exact shape of the unity-gain transfer function which determines the transient behavior of the amplifier and fixes the relation between open-loop and closed-loop bandwidth. After discussing the effects of slew rate and pole-zero doublets on the transient response, we are ready to consider complete amplifier realizations in chapter 5.

Limitations of the dc-gain and the bandwidth are presented in section 4.2 and section 4.3, respectively. Thereafter, the basics of frequency compensation are discussed in section 4.4, and frequency compensation techniques for two-stage amplifiers are treated in section 4.5. Thereafter, nested-Miller compensation, cascoded-nested-Miller compensation, and mirrored-nested-Miller compensation, which are techniques for compensating three-stage amplifiers, are presented in section 4.6. Dynamic distortion and settling time are analyzed in section 4.7 and section 4.8, respectively. Designing for programmability is discussed in section 4.9 and, finally, conclusions are drawn in section 4.10.

4.2 DC-gain limitations

There are two ways to achieve sufficient gain. First of all, gain can be achieved by cascading a number of stages. Secondly, cascoding can be applied to improve the gain of single stages.

A cascaded topology is shown in Fig. 4-1a in bipolar technology, and in Fig. 4-1b in CMOS technology. The gain A_0 of the bipolar cascaded topology is given by

$$A_0 = g_{mn}\beta_F^{N-1}R_L = \frac{qI_N}{kT}\beta_F^{N-1}R_L \tag{4-1}$$

where N is the number of stages, g_{mN} is the transconductance of the input stage, and I_N is the bias current of the input stage. The gain of a single stage topology is equal to the product of the transconductance of the input stage and the load resistance, while each added intermediate stage contributes a factor β_F to the overall gain. The result is a little bit disappointing in view of the large possible gain of a single bipolar stage. However, due to the low input impedance of bipolar transistors the performance of a cascaded string of transistors is mediocre. Further, we note that the gain is proportional to the bias

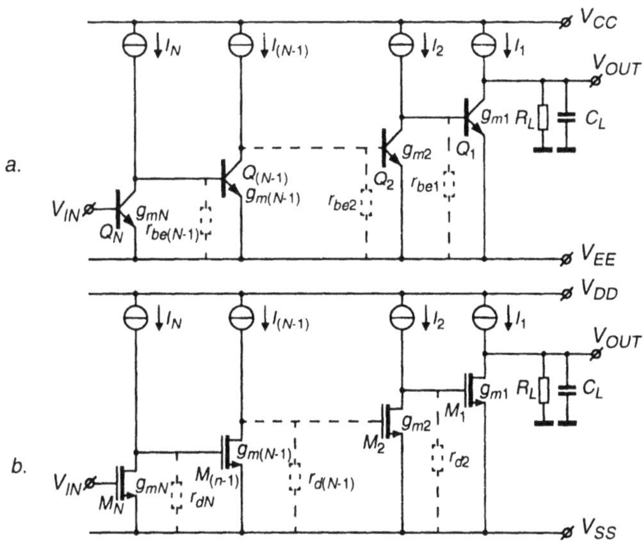

Fig. 4-1 *Achieving gain using a cascaded structure in bipolar technology (a) and CMOS technology (b).*

current of the input stage. Using equation (4-1) an expression for the required number of stages can be calculated

$$N = 1 + \frac{\ln\left(\frac{A_0}{g_{mN}R_L}\right)}{\ln\beta_F} \tag{4-2}$$

If, for example, the product of g_{mN} and R_L is about one and β_F is 100, a gain of 80 dB can only be attained using at least five stages. In CMOS in strong inversion, the gain of a cascaded circuit is given by

$$A_0 = g_{m1}(g_m r_d)^{N-1} R_L = \sqrt{2\mu C_{ox}\frac{W_1}{L_1}I_1}\left(\frac{3\mu C_{gs}}{I_i}\right)^{\left(\frac{N-1}{2}\right)} R_L \tag{4-3}$$

The gain is proportional to the square root of the current of the output stage which is in accordance with the biasing requirements for Miller-compensated circuits. On the other hand, the appearance of C_{gs} gives an additional presentiment of the conflicting requirements for gain and bandwidth. Due to the high input impedance of CMOS transistors, the contribution of each intermediate stage to the overall gain is equal to the maximum gain of a single stage.

The high input impedance also enables high gain using cascodes. In fact, the contribution of a cascode to the gain is of the order of a cascaded stage. Because the frequency of the pole introduced by the cascode is very high, cascoding is an effective way of combining high gain with high bandwidth in CMOS technology. An example of a multi-stage topology with M cascodes M_{C1}-M_{CM} is depicted in Fig. 4-2. Using cascodes the required

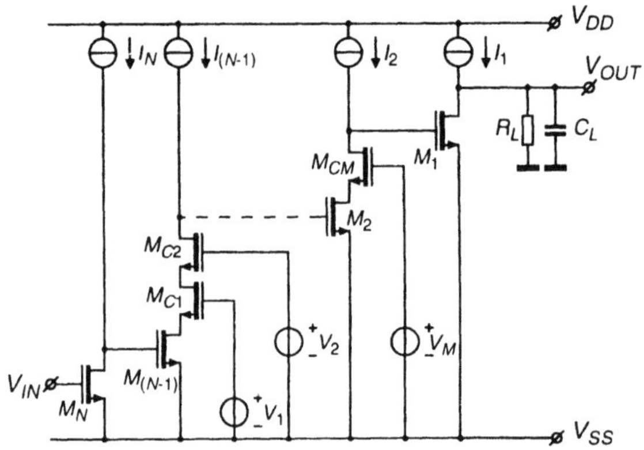

Fig. 4-2 *CMOS multi-stage amplifier with cascodes in order to combine high gain with a low number of stages.*

number of stages N for a given amount of gain is

$$N = 1 - M + \frac{\ln\left(\frac{A_0}{g_{m1}R_L}\right)}{\ln\left(g_m r_d\right)} \tag{4-4}$$

where M is the number of cascodes. Placing several cascodes on top of each other, as applied in Fig. 4-2, is not possible at low supply voltages. By using gain boosting [1] however, high gain can be obtained using a single cascode. A simplified circuit diagram of a two-stage amplifier with a gain-boosted cascode is shown in Fig. 4-3. The output impedance of cascode M_{21} is boosted by amplifier stage M_{22}. In this way a very high impedance is created at the gate of output stage M_1. If necessary, the gain can be further increased by cascoding M_{22}, or even by using a gain-boosted cascode as cascode for

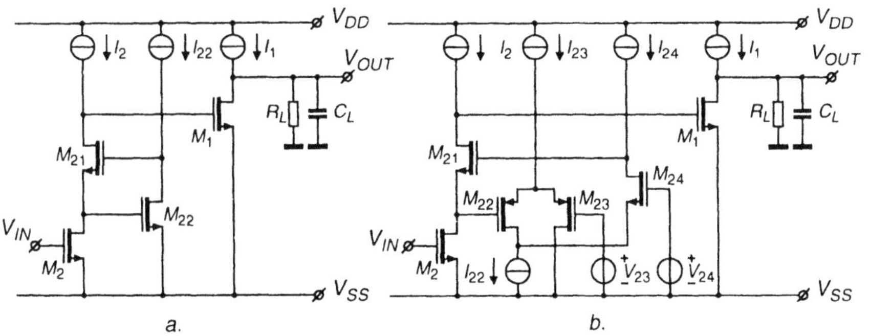

Fig. 4-3 *Principle of two-stage CMOS amplifier with gain boosting (a) and low-voltage version (b).*

M_{22}. The gain-boosting principle can be adapted for low-voltage operation by using a folded cascode stage, as shown in Fig. 4-3b [2], consisting of PMOS differential stage M_{22}, M_{23} and NMOS folded cascode M_{24}. Thus, a very high gain can be created in CMOS technology at low supply voltages by using only two stages. At very low supply voltages, folded cascodes may be used but cannot contribute to the gain because it is not allowed to use cascoded current sources on top of the gate-source voltage of the output transistor, as is discussed in more detail in section 5.3.3. In that case cascading is the only option.

In bipolar technology the gain can be increased by improving the input impedance of a stage. A well-known example is the Darlington stage Q_1, Q_2 as used in the circuit shown in Fig. 4-4. If the output impedance of Q_3 is sufficiently high, the contribution to the

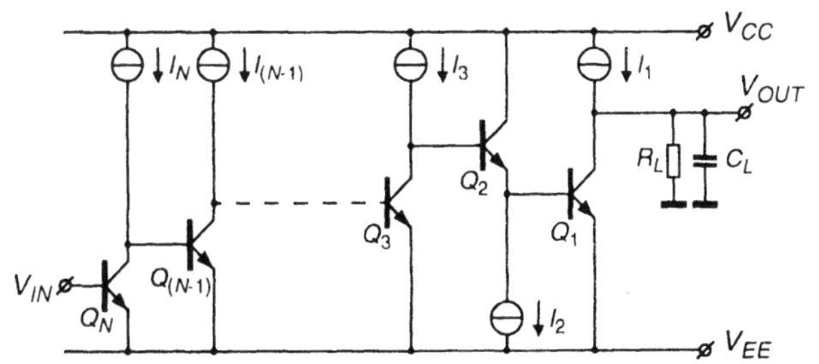

Fig. 4-4 *Multi-stage amplifier in bipolar technology using Darlington transistor in output stage.*

overall gain of Q_2 is comparable to the gain enhancement of a common-emitter stage.

Since the frequency of the additional pole of the Darlington at the emitter of Q_2 is very high, the circuit initially behaves as a two-stage circuit. However, if the Darlington is used in a class-AB stage and the output current is increased, peaking occurs. This can be prevented by applying nested-Miller compensation to the Darlington stage. Therefore, if applied in a class-AB output stage, the Darlington still behaves as a cascade of transistor stages. Another drawback of the Darlington is that a relatively large supply voltage is required of at least two base-emitter voltages and a saturation voltage. Other techniques for improving the input impedance of bipolar transistors use base-current compensation. Base-current compensation employs a positive feedback loop of which the gain should be smaller than unity. The improvement using positive feedback is limited and the positive feedback results in high-frequency stability problems.

4.3 Bandwidth limitations

The transfer function of the bipolar cascaded amplifier as shown in Fig. 4-5a, can be approximated for high frequencies by noting that the resistive impedances can be disregarded, yielding

$$\frac{v_{out}}{v_{in}} = \frac{\prod_{i=1}^{N} g_{m,i}}{s^N C_L \prod_{j=1}^{N-1} C_{be,j}} = \frac{C_{be1}}{s^N C_L} \prod_{i=1}^{N} \frac{g_{m,i}}{C_{be,i}} \tag{4-5}$$

The unity-gain frequency of equation (4-5) is a measure for the maximum achievable bandwidth ω_n which is given by

$$\omega_n = \sqrt[N]{\frac{C_{be1}}{C_L}} \prod_{i=1}^{N} \sqrt[N]{\frac{g_{m,i}}{C_{be,i}}} \tag{4-6}$$

The bandwidth ω_n is maximized by biasing the same current in all stages. If we further assume that all the base-emitter capacitances are equal, we can write

$$\omega_n = \frac{qI_{SUP}}{NkT} \sqrt[N]{\frac{C_{be}}{C_L}} \tag{4-7}$$

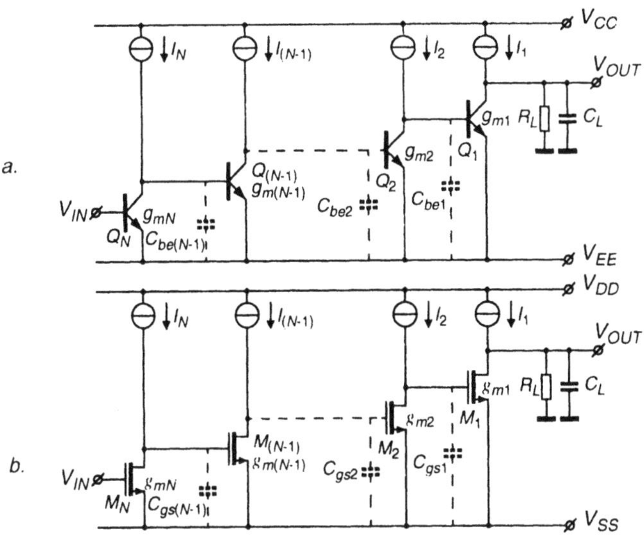

Fig. 4-5 *Cascaded amplifier in bipolar technology (a) and CMOS technology (b).*

where each stage is biased by a current I_{SUP}/N. Differentiating equation (4-7), with respect to the number of stages N, the optimum number of stages can be found, given by

$$N = \ln\left(\frac{C_L}{C_{be}}\right) \tag{4-8}$$

For the CMOS cascaded amplifier shown in Fig. 4-5b, we find the bandwidth as

$$\omega_n = \frac{1}{L}\sqrt{\frac{3\mu I}{NC_{gs}}}\sqrt[N]{\frac{C_{gs}}{C_L}} \tag{4-9}$$

and the optimum number of stages is given by

$$N = 2\ln\left(\frac{C_L}{C_{gs}}\right) \tag{4-10}$$

Further, the amplifiers require frequency compensation which generally reduces the bandwidth and affects the above given findings. If a compensation technique is used that is ideal regarding maximizing the bandwidth, the findings are correct. An example of such a compensation technique is pole-zero cancellation, of which the basic two-stage form is discussed in section 4.5.1. If we design a bipolar amplifier using pole-zero cancellation for a load capacitor equal to $100C_{be}$, we need four to five stages to maximize the bandwidth. In order to obtain Butterworth pole positions after unity-gain feedback, the non-dominant poles should be placed in complex positions. This requires parallel networks which are not easy to create. Further, due to the appearance of pole-zero doublets which are not accurately matched, this technique cannot be used in a general purpose amplifiers. The only compensation technique that is insensitive to parameter changes and does not employ pole-zero cancellation is nested-Miller compensation. However, when nested-Miller compensation is used the bandwidth reduces due to the insertion of the Miller capacitor. In addition, the bandwidth reduces as a function of the number of stages.

The actual performance of the amplifier is not only determined by the compensation technique and the number of stages, but also by the type of transfer function. In most amplfier designs the Butterworth transfer function is used. But, why is the Butterworth functions so often used? To answer this question several types of second-order transfer function are compared to a first-order single-pole function in Table 4-1 [3]. In the table ζ is the damping ratio which determines the shape of the transient response as we will see in a moment. The damping ratio has a direct relation with the position of the second pole

$$\omega_2 = 4\zeta^2\omega_0 \qquad (4\text{-}11)$$

and the phase margin

$$\varphi_m = \operatorname{atan}(4\zeta^2) \qquad (4\text{-}12)$$

and is a measure for the stability of the system. Designing amplifiers with a damping ratio close to unity yields very stable systems but costs a lot of bias current. Designing, for example, a transfer function according to the Bessel function requires the second pole to be at a three times higher frequency than when the amplifier is designed according to the half critical damping function. Therefore, a factor three more power is required to implement the Bessel response with the same unity-gain frequency as the half-critical-damping response. About the same factor is found if we consider the closed loop $-3dB$ frequency ω_{-3dB}. Amplifiers designed according to the Butterworth function obtain a reasonable compromise between power consumption and phase margin. The phase margin of more than 60° allows for large variations in components value, resulting in a guaranteed stable amplifier.

Table 4-1 *Frequency-domain parameters of second-order functions compared to single pole system*

Function type	Closed-loop parameters					Openloop parameters	
	ζ	ω_n	ω_{-3dB}			ω_0	φ_m [°]
Half critical damping	$\frac{1}{2}$	ω_0	$1.272\omega_n$	$1.272\omega_0$	$1.272\omega_2$	ω_2	45
Butterworth	$\frac{1}{2}\sqrt{2}$	$\sqrt{2}\,\omega_0$	ω_n	$1.414\omega_0$	$0.707\omega_2$	$\frac{1}{2}\omega_2$	63.4
Bessel	$\frac{1}{2}\sqrt{3}$	$\sqrt{3}\,\omega_0$	$0.786\omega_n$	$1.362\omega_0$	$0.454\omega_2$	$\frac{1}{3}\omega_2$	71.6
Coincident poles	1	$2\omega_0$	$0.644\omega_n$	$1.287\omega_0$	$0.322\omega_2$	$\frac{1}{4}\omega_2$	76.0
Single pole		ω_0	ω_n	ω_0			90

4.4 Frequency compensation

In the previous sections we have discussed dc-gain limitations and bandwidth limitations. We found that a certain number of stages are required to obtain sufficient gain, depending on the possibility to employ cascodes to boost the gain of single stages. With regard to bandwidth, ideally, an optimum number of stages can be found. In practice, however, the relation between bandwidth and number of stages is heavily influenced by the applied frequency compensation techniques. It is therefore time to consider the basics of frequency compensation in this section. Afterwards, the implementation and optimalization of frequency compensation in two-stage and three-stage amplifiers is presented in subsequent sections.

In order to investigate the stability of feedback systems, the time dependency has to be introduced resulting in a set of nonlinear differential equations that completely describe the system. However, solving such a set of differential equations is not possible. Fortunately, by assuming the changes of signals and parameters to be relatively small, we can approximate the system by a set of linear differential equations in the same way as the small signal approximation of dc parameters. Large signal behavior can later on be treated by studying the influence of changes of the signal and transistor parameters. Since linear differential equations have been studied extensively, there are a lot of tools available to deal with the set of equations. The most convenient way is to introduce the

Laplace transform which transforms the differential equations into a set of ordinary equations which are relatively easy to solve. Using the Laplace transform the fundamental feedback [4] equation can be written as

$$H(s) = \frac{A(s)}{1 + A(s)\beta(s)} \tag{4-13}$$

Introducing the stability requirement that the response of the circuit must be bounded yields that the poles are not allowed to be positioned in the right-half plane. The poles are determined by the characteristic equation $1+A(s)\beta(s)$. While the poles and zeros of $A(s)$ are usually available in a decomposed form, the solution of the characteristic polynomial is not available and usually difficult to obtain. In the design of operational amplifiers, we do not even know the feedback factor $\beta(s)$. It is therefore convenient to be able to predict the stability of the closed-loop system $H(s)$ without having to calculate the characteristic equation base on the positions of the poles and zeros of $A(s)$. This is possible using the Nyquist criterion [5]. The description and application of the Nyquist criterion itself is beyond the scope of this work and can be found in other text-books [6,7].

We use the derived measures phase margin and gain margin. Phase margin is the difference between 180° and the phase lag at the point were the loop gain is equal to unity. If the phase lag reaches 180° before the amplitude response is below unity gain, the feedback reverses and the system breaks into oscillation. Thus, the phase margin can be regarded as the amount of phase lag that can be added before the system turns into an oscillator. The gain margin is the inverse of the gain of $A(s)$ at the frequency where the phase characteristic crosses 180°, and can be regarded as the amount of gain that can be added before the circuit starts oscillating. Practically, phase margin is mainly used to quantify the stability of feedback systems. Most circuits are designed having a phase margin of the order of 60° corresponding to the Butterworth transfer function. Gain margin depends on the order of the system. A second order system theoretically has infinite gain margin unless we take high order poles and right-half-plane zeros into account. However, gain margin is an important measure to correctly judge circuits that show excellent phase margin, but hardly have any gain margin. Examples are circuits with complex poles and circuits with one or more zeros just below unity gain .

As an example, Fig. 4-6 shows a three-stage amplifier consisting of input stage M_{31}, M_{32}, intermediate stage M_{21}, M_{22}, and output stage M_1. The block $\beta(s)$ contains a feedback network that feeds the output voltage, or part of the output voltage, back to input transistor M_{31}. The worst case situation is a unity-gain feedback when the whole output voltage is fed back to the input stage. The amplifier has three dominant poles caused by capacitors C_2, C_1, and C_L, resulting in an open-loop Bode plot as indicated in

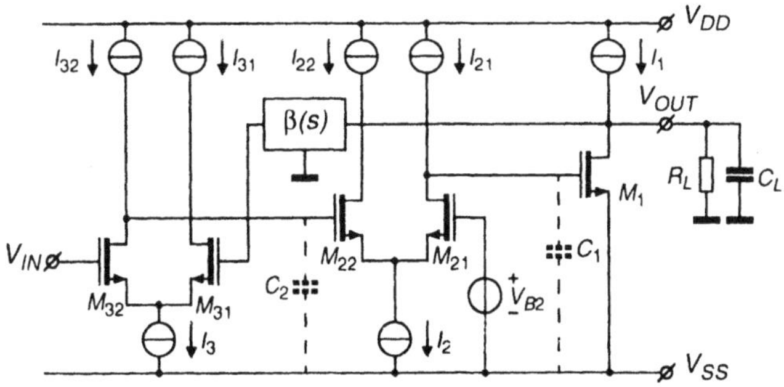

Fig. 4-6 *Simplified circuit diagram of an example of a three-stage amplifier with feedback β(s).*

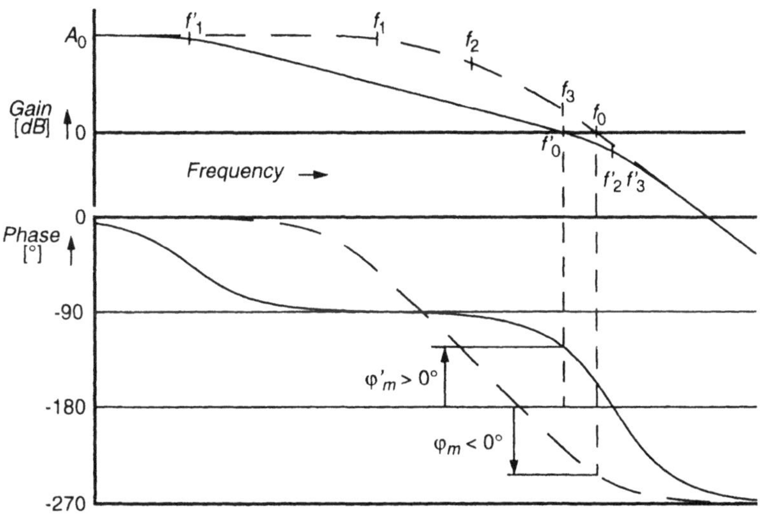

Fig. 4-7 *Bode diagram of three-stage amplifier. Dashed line indicates frequency response without frequency compensation. Solid line indicates frequency response after proper frequency compensation techniques have been applied.*

Fig. 4-7 by the dashed line. Clearly, the phase margin φ_m of the uncompensated circuit is negative, indicating that a unity gain feedback will result in oscillations. One way of obtaining a stable frequency response is by shifting one of the poles into a dominant position while shifting the other poles to non-dominant locations. The resulting compensated characteristic is plotted in Fig. 4-7 by the solid lines. Now, the phase margin is much larger than 0°, and unity-gain feedback is allowed. For reasons of clarity, the gain margin has not been indicated in Fig. 4-7. However, it is easy to verify that the gain margin of the uncompensated characteristic is much smaller than 0 *dB*, while the gain margin of the compensated system is much larger than 0 *dB*. Shaping the frequency response in such a way that a single dominant pole is obtained is not the only way of frequency compensation. Using the Bode relations [4] it is possible to define a frequency characteristic that achieves maximum possible gain at each frequency, while having a certain amount of phase margin. The resulting so-called maximum feedback amplifier [6] requires a large number of capacitors and inductors to implement the required zeros and poles and is therefore not suited to integrated technology [7].

A final frequency shape of interest is found in conditionally stable amplifiers [7]. In such amplifiers, a couple of zeros are introduced so that a first order characteristic is obtained in a certain, but limited, part of the frequency characteristic. Thus, higher gain is obtained in the mid-frequency band, but only a small range of feedback factors result in stable amplifiers. Reliable circuits in integrated technology have been demonstrated recently [7], but due to the limited range of feedback factors this technique is not suited to general purpose operational amplifiers. Hence, the only characteristic that can be used in general purpose integrated amplifiers is the single-dominant pole characteristic, which yields phase margin over the whole frequency range up to unity gain so that any feedback factor can be applied.

The rest of this chapter is devoted to the implementation of such a characteristic in amplifiers consisting of an arbitrary number of stages. But first we briefly consider what exact positioning of the poles is desirable. The desirable stability forms a compromise between power efficiency, settling of the step response, and overshoot of the step response. Closed-loop amplifier characteristics based on Butterworth polynomials obtain a good compromise for general purpose amplifiers as discussed in section 4.3 and section 4.8. In this section we only briefly discuss the Butterworth characteristics. An overview of Butterworth polynomials up to the fourth-order equation is given in Table 4-2. Also, the equations and pole positions of the corresponding open-loop systems before unity-gain feedback is applied are indicated. The poles of the unity-gain Butterworth system are placed on a semi-circle and the angle α between the poles is given by a simple rule

$$\alpha = \frac{180°}{n} \qquad (4\text{-}14)$$

It is an important advantage that the pole positions are given by such a simple geometric rule. Pole positions of other polynomials cannot be described in such a simple way and require manipulation of the polynomial coefficients, at least for higher order polynomials. Another advantage appears if we consider how to obtain the open-loop pole configurations as given by Table 4-2, of the higher order systems. For, example, the third-order open-loop system can be created by adding a dominant pole to a closed-loop second-order system. This observation enables the simple design of nested compensation

Table 4-2 *Butterworth pole positions and pole positions of corresponding open-loop systems*

n	Butterworth pole-positions	open-loop system
1	$s+\omega_n$ 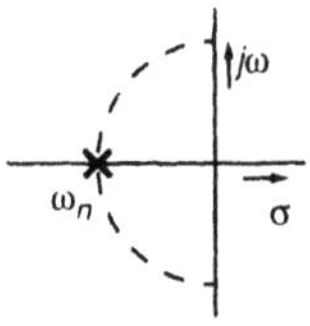	$s+\frac{\omega_n}{A_0}$ 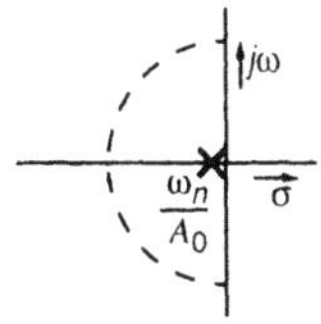
2	$s^2+\sqrt{2}\omega_n s+\omega_n^2$ 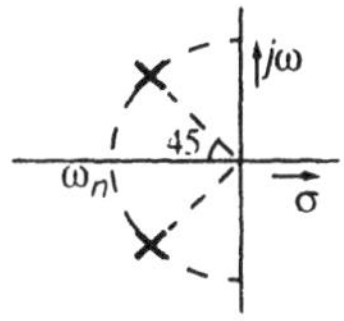	$\left(s+\frac{1}{2}\sqrt{2}\frac{\omega_n}{A_0}\right)\left(s+\sqrt{2}\omega_n\right)$ 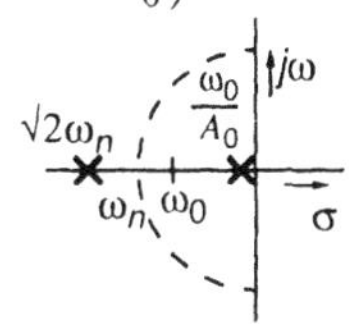
3	$s^3+2\omega_n s^2+2\omega_n^2+\omega_n^3$ 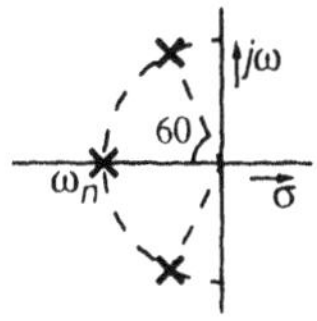	$\left(s+\frac{1}{2}\frac{\omega_n}{A_0}\right)\left(s^2+2\omega_n s+2\omega_n^2\right)$ 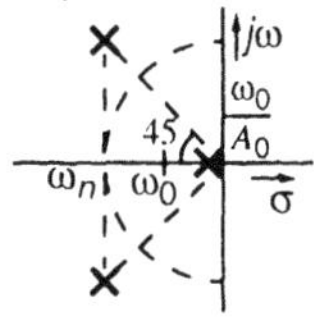
4	$s^4+2.61\omega_n s^3+3.41\omega_n^2 s^2$ $+2.61\omega_n^3 s+\omega_n^4$ 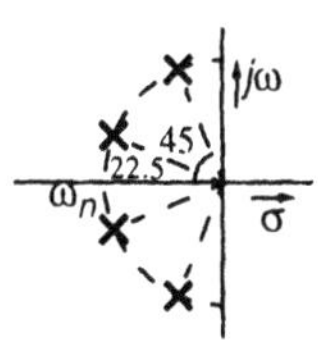	$\left(s+0.38\frac{\omega_n}{A_0}\right)\left(s^3+2.61\omega_0 s^2+3.41\omega_0^2 s+3.24\omega_0^3\right)$ 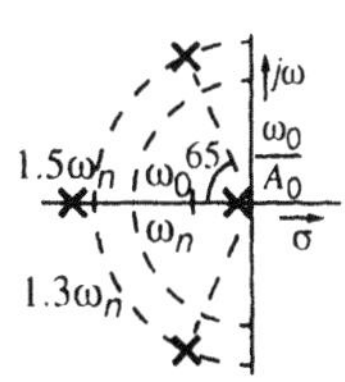

structures. Again, this property is not found in other pole-position schemes, and, even with Butterworth polynomials, it is only valid for lower order systems. For example, the three dominant poles of the fourth-order system are not exactly at the locations of the third-order closed-loop system. Some other important properties of Butterworth equations are listed in Table 4-3. The -3*dB* bandwidth of the closed-loop system is

Table 4-3 *Closed-loop and open-loop parameters of systems with unity-gain pole positions in Butterworth location (The values in parentheses are obtained with non-dominant poles at ten times the unity-gain frequency. For the first order system two poles have been added and for the second order systems one pole has been added.).*

n	ω_{-3dB}	ω_0	φ_m [°]	$\omega_{\varphi=180}$	γ_m [*dB*]
1	ω_n	ω_n	90(78.6)	∞ ($10\omega_n$)	∞(26.0)
2	ω_n	$\frac{1}{2}\sqrt{2}\omega_n$	63.4(57.7)	∞ ($4.5\omega_n$)	∞(21.6)
3	ω_n	$\frac{1}{2}\omega_n$	60.3	$\sqrt{2}\omega_n$	12.6
4	ω_n	$0.38\omega_n$	60.7	ω_n	7.66

independent of the order and equal to the radius of the semi-circle. Also, except for the first-order systems, the phase margin is nearly independent of the order. This suggests a further simplification of the nested design procedure since we can use the phase margin as a criterion, and do not need to calculate and manipulate the coefficients or poles of the system equations. Finally, we note that the unity-gain frequency and the gain margin reduce as a function of the order n. The reduction in unity-gain frequency is regained when unity-gain feedback is applied, since the -3*dB* frequency is independent of n. However, because the opamp is not always used in unity-gain configuration, this factor is often lost. It is therefore advantageous to keep the order of the system as small as possible. Multipath-nested-Miller compensation discussed in section 4.6.4 is a helpful tool to reduce the order of the system.

Concluding, pole-positions based on Butterworth polynomials have several advantages and will therefore be used in this chapter to define the desired pole-positions of the amplifiers to be frequency compensated. The advantages of Butterworth polynomials include: easy rules for the placement of the closed-loop poles, closed-loop -3*dB* bandwidth that is independent of the order of the system, closed-loop poles of order n form the basis for the placement of the open-loop poles of the system of order n+1 enabling a nested compensation technique, and, finally, the phase margin is independent of the order and equal to about 60°. All these advantages considerably reduce the complexity of the design of n-stage amplifiers. In fact, the design problem of an n-stage amplifier is transformed into the design of n-1 two-stage amplifiers which all require a phase-margin of 60°. We now consider actual amplifier configurations starting with two-stage amplifiers.

4.5 Two-stage amplifier optimalization

Two-stage amplifier configurations are very common in CMOS where cascoding techniques can be applied to boost the gain. In bipolar technology, and at very low supply voltages, more stages are required. Compensation techniques for circuits consisting of more stages are based on the two-stage Miller compensation and, therefore, the techniques discussed in this section are of importance to all amplifier designs.

Although, in principle, resistive techniques can be applied to obtain a stable frequency response such as resistive broadbanding, these techniques severely limit the gain and are not suitable for application in general purpose high-gain amplifiers. These techniques are therefore not discussed.

First, parallel compensation is discussed. Although the application of parallel compensation is limited because it relies on pole-zero cancellation and requires relatively large compensation capacitor values, it serves as an important reference because the highest possible bandwidth can be obtained using this compensation scheme. Afterwards, Miller compensation is treated. Miller compensation results in a smaller bandwidth, but the value of the compensation capacitor is much smaller and no pole-zero doublets are introduced.

4.5.1 Parallel compensation

As indicated in Fig. 4-8, where a circuit diagram of a two-stage amplifier is shown, the parallel compensation technique consists of a parallel network consisting of a resistor R_P and a capacitor C_P that is connected to the internal node of the amplifier. Capacitor C_P increases the capacitive load at the node resulting in a dominant pole. Without inserting R_P this would result in a large bandwidth reduction. By inserting R_P, a zero is created at frequency $1/R_PC_P$ which should be placed at the frequency of the second pole. The pole is then replaced by a pole at a higher frequency so that frequency compensation is obtained without bandwidth reduction. The mechanism is illustrated by the bode plot as shown in Fig. 4-8, where the uncompensated and compensated characteristic are

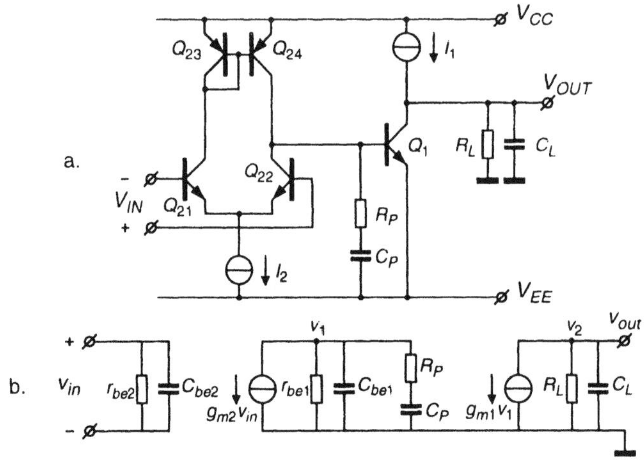

Fig. 4-8 *Two-stage amplifier in bipolar technology with parallel compensation at the base of the output transistor (a) and small-signal equivalent circuit (c).*

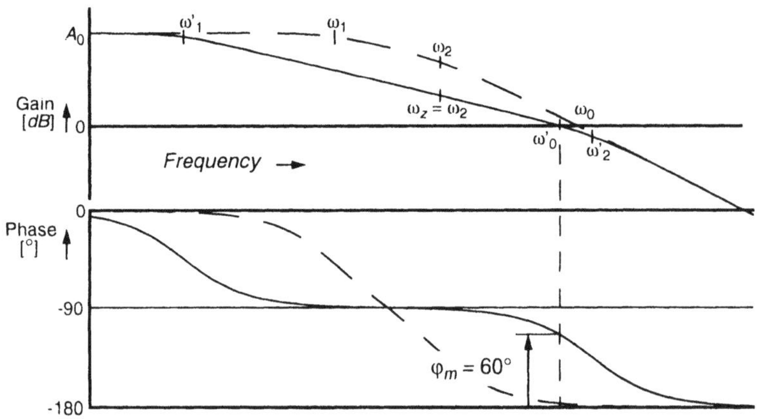

Fig. 4-9 *Frequency response of two-stage amplifier with parallel compensation. Dashed line indicates uncompensated characteristic, solid line indicates compensated characteristic.*

plotted together. Because of the cancellation of a pole by a zero, the technique is also called pole-zero cancellation.

A drawback of parallel compensation is that the condition for pole-zero cancellation as given by

$$R_P C_P = R_L C_L \tag{4-15}$$

is difficult to satisfy in general purpose amplifiers, because the load is usually not known. Therefore, application of this technique results in a pole-zero doublet with a large spacing between the pole and the zero which increases the settling time of the amplifier.

Another disadvantage of parallel compensation becomes clear if we consider the open-loop unity-gain frequency

$$\omega_0 = \frac{g_{m2} g_{m1} R_L}{C_P} \tag{4-16}$$

Comparing the unity-gain frequency of parallel compensation to the unity-gain frequency of Miller compensation, as given by equation (4-22), we conclude that the required parallel capacitor is a factor of $g_{m1} R_L$ larger than the required Miller capacitor for the same bandwidth.

Nevertheless, this compensation scheme is important because it obtains the highest possible bandwidth that can be achieved using a two-stage amplifier. The bandwidth is given by

$$\omega_n = \sqrt{\frac{g_{m1} g_{m2}}{C_L C_{be1}}} \tag{4-17}$$

The value of ω_n can even be higher than that of a single stage amplifier. It is useful as a reference for other compensation schemes and as a limit for what is maximally possible.

It is interesting to write the bandwidth as a function of the supply current and consider the amount of current I_2 that should be biased in input stage, and the current I_1 that should be used in the output stage. If α is the fraction of the total supply current I_{SUP} used to bias the input stage, and the remaining current is used in the output stage, we find

$$\omega_n = \frac{1}{2}\sqrt{2}\frac{1}{V_T}\sqrt{\frac{I_1 I_2}{C_L C_{be1}}} = \frac{1}{2}\sqrt{2}\frac{I_{SUP}}{V_T}\sqrt{\frac{\alpha(1-\alpha)}{C_L C_{be1}}} \tag{4-18}$$

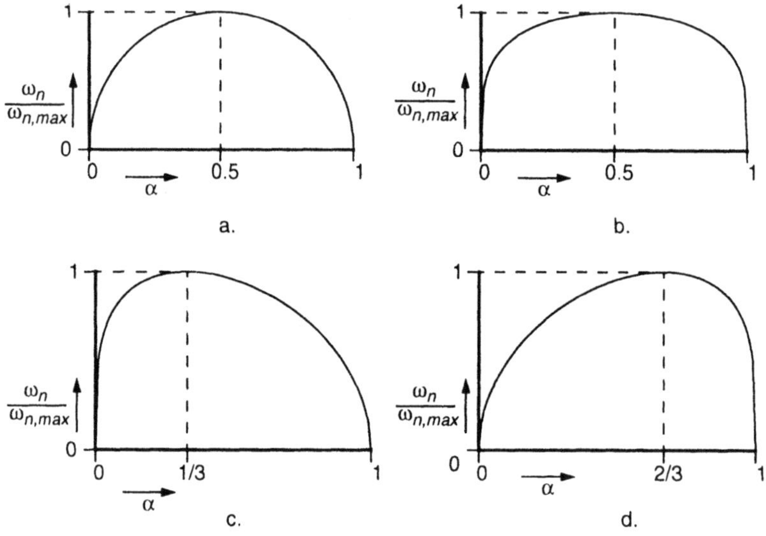

Fig. 4-10 *Bandwidth ω_n as a function of current division parameter α in bipolar technology (a), CMOS technology (b), BiCMOS technology with bipolar output stage (c) and BiCMOS technology with bipolar input stage (d).*

The bandwidth as a function of α is plotted in Fig. 4-10a. As we see, a relatively flat optimum is found for α=0.5. If we bias the amplifier around this value, the maximum bandwidth

$$\omega_{n,\,max} = \frac{1}{4}\sqrt{2}\frac{I_{SUP}}{V_T}\frac{1}{\sqrt{C_L C_{be1}}} \tag{4-19}$$

is achieved. In the above calculation it has been assumed that the value of C_{be1} is constant. In practical situations, the base-emitter capacitance changes as a function of the collector current. This effect can be modeled by a placing C_{be1} by an approximation that is linearly dependent on the bias current and results in a slight change of the shape of the bandwidth plot. In CMOS technology in strong inversion, similar equations can be derived. In CMOS the maximum bandwidth is also found for α=0.5. The bandwidth as a function of α in strong inversion is plotted in Fig. 4-10b. Because the transconductance in strong inversion is proportional to the square root of the current, the maximum around

α=0.5 is flatter than in bipolar technology. The maximum bandwidth in CMOS in strong inversion is equal to

$$\omega_{n,\,max} = \frac{1}{L}\sqrt{\frac{3\mu I_{SUP}}{2C_L}}\sqrt[4]{\frac{W_2}{2W_1}} \qquad (4\text{-}20)$$

where it has been assumed that the length L of the input transistor is equal to the length of the output transistor. Obviously, the length of the transistors should be as small as possible. Further, the width W_1 of the output transistor should be small compared to the width W_2 of the input transistor because increasing W_1 increases the gate-source capacitance of the output stage, assuming the gate-source capacitor is much larger than the gate-source capacitance of the input stage. In BiCMOS technology using different types of transistors or using CMOS technology with one stage biased in strong inversion and the other one in weak inversion, the optimum is not at α=0.5. If a bipolar output stage is used or a CMOS stage biased in weak inversion, the optimum is at $\alpha={}^1/_3$ as shown in Fig. 4-10c. If the input stage is implemented using a bipolar transistor or a CMOS transistor operating in weak inversion, the optimum is around $\alpha={}^2/_3$ as depicted in Fig. 4-10d.

4.5.2 Miller compensation

Due to the pole-zero doublet and the relatively large value of the parallel capacitor, parallel compensation, which is discussed in the previous section, is not suited to general purpose amplifiers and Miller compensation should be used.

Miller compensation is a technique where feedback is used to enhance the operation of the compensation capacitor. Therefore, contrary to parallel compensation, the value of the compensation capacitor is relatively small and the behavior of the compensation is much less sensitive to changes in process and other parameters. First, the fundamentals of Miller compensation are discussed and the behavior under various operating conditions is considered. It is shown that the price to be paid for the improved accuracy is that the achievable bandwidth is smaller. Afterwards, a technique that improves the bandwidth of basic Miller compensation using a cascode in the Miller loop is discussed in section 4.5.4.

A two-stage bipolar amplifier with Miller compensation [8,7] is depicted in Fig.4-11a. The basic technique consists of a Miller capacitor C_M inserted across the base-collector junction of the output transistor. Due to the gain of the output transistor, the capacitor determines the frequency response in the mid-frequency range. Assuming the gain of the

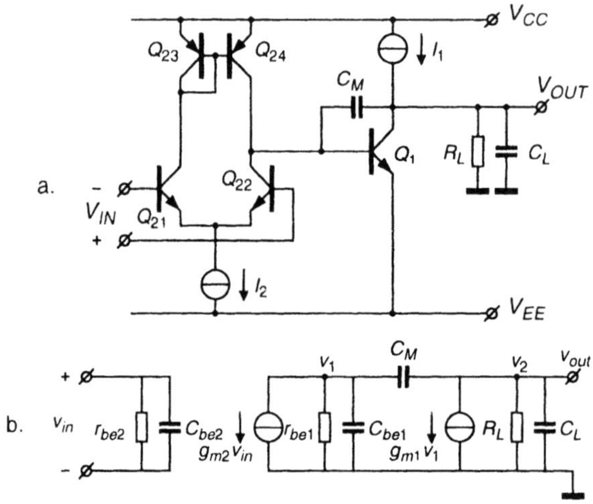

Fig. 4-11 *Two-stage amplifier in bipolar technology with Miller compensation (a) and small signal equivalent circuit (b).*

output stage is high, the transfer function of the output stage in the mid-frequency range is given by

$$\frac{v_{out}}{v_{in}} = \frac{g_{m2}}{sC_M} \tag{4-21}$$

from which the unity-gain frequency can easily be calculated as

$$\omega_0 = \frac{g_{m2}}{C_M} \tag{4-22}$$

Comparing equation (4-22) to the unity-gain frequency of parallel compensation as given by equation (4-16), it turns out that the Miller capacitor C_M is a factor $g_{m1}R_L$ smaller than the parallel capacitor for the same bandwidth, which is an important advantage. Combining equation (4-22) with the dc-gain, the dominant pole can be calculated

$$\omega_1 = -\frac{1}{g_{m1}r_{be1}R_LC_M} \tag{4-23}$$

The second pole can be found by inspection from the circuit while assuming that the Miller capacitor can be considered as a short circuit for high frequencies. In that case, a pole remains which is determined by C_{be1}, C_L, R_L and the impedance of the diode-connected output transistor which is equal to $^1/g_{m1}$. All the elements are connected in parallel so that the second pole is easily calculated as

$$\omega_2 = -\frac{g_{m1} + \frac{1}{R_L}}{C_L + C_{be1}} = -\frac{g_{m1}}{C_L}\left(\frac{1 + \frac{1}{g_{m1}R_L}}{1 + \frac{C_{be1}}{C_L}}\right) \tag{4-24}$$

Normally, the voltage gain $g_{m1}R_L$ of the output stage is large and the base-emitter capacitance is much smaller than the load capacitor, so that the second pole is mainly determined by g_{m1} and C_L. Using equation (4-24) and equation (4-22), the value of the Miller capacitor can be calculated in order to obtain a certain amount of phase margin. If we require the poles of the unity-gain connected amplifier to be in Butterworth position, the value of the Miller capacitor is

$$C_M = 2\frac{g_{m2}}{g_{m1}}C_L \tag{4-25}$$

Thus, the required value of C_M is determined by the load capacitor and the ratio between the transconductance of the input stage and the transconductance of the output stage. Since the ratio of transconductances is well controlled in IC technology and the absolute value of a good-quality capacitor is also well controlled, this means that stability can be guaranteed up to a maximum load capacitance over process variations, temperature, and other parameters which gives a major advantage over the pole-zero cancellation technique. The resulting frequency response of the Miller compensated two-stage amplifier is depicted in Fig. 4-12, illustrating the pole-splitting behavior of the Miller capacitor which results in sufficient phase margin. Also, the plot shows that at high frequencies the characteristic of the compensated circuit does not match with the original uncompensated frequency response. Therefore, the second pole is placed at a lower frequency than the second pole of a parallel compensated circuit. This bandwidth reduction can be quantified by calculating ω_n of the Miller compensated circuit. The simplest way to calculate ω_n is by calculating the transfer function for higher frequencies using the small-signal equivalent circuit shown in Fig.4-11a, while the resistors can be

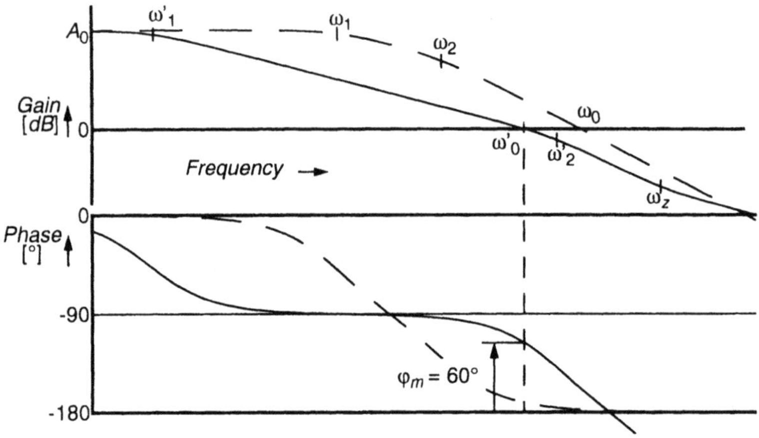

Fig. 4-12 *Frequency response of two-stage amplifier with Miller compensation. Dashed line indicates uncompensated characteristic, solid line indicates compensated characteristic.*

removed for these high frequencies. The unity-gain frequency of the high-frequency transfer function yields

$$\omega_n = \sqrt{\frac{g_{m1}g_{m2}}{C_{be1}C_L}}\frac{1}{\sqrt{1+\frac{C_M}{C_L}+\frac{C_M}{C_{be1}}}} \qquad (4\text{-}26)$$

where the second factor represents the bandwidth reduction caused by the Miller capacitor. This reduction is caused by the capacitive loading by C_M of the output node and the base connection of the output transistor. Since C_L is usually much larger than C_{be1}, the bandwidth reduction is mainly caused by the ratio C_M over C_{be1}. This suggests that we should design the circuit in such a way that C_M is as small as possible. However, there is a certain optimum. A small C_M causes g_{m2} to be small which reduces the bandwidth. The optimum can be found by combining equation (4-26) with equation (4-25), and by introducing the current division parameter α in the same way as with parallel compensation, yielding [9]

$$\omega_n = \frac{1}{2}\sqrt{2}\frac{I_{SUP}}{V_T}\sqrt{\frac{\alpha(1-\alpha)}{C_{be1}C_L + \frac{\alpha}{1-\alpha}C_L(C_{be1}+C_L)}} \tag{4-27}$$

The bandwidth ω_n as a function of α is plotted in Fig. 4-13 for several load-capacitor values. As a reference the plot of parallel compensation is indicated by the dashed line.

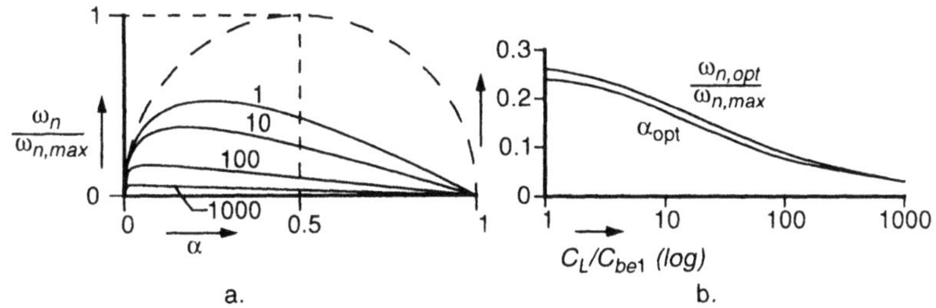

Fig. 4-13 *Bandwidth ω_n as a function of current division parameter α in bipolar technology of a Miller compensated two-stage amplifier for several values of C_L/C_{be1} (a) and the optimal α and resulting bandwidth $\omega_{n,opt}$ as a function of C_L/C_{be1} (b).*

For low capacitive loads the input bias current should be equal to 25% of the total bias current in order to attain the maximum bandwidth. For high capacitive loads nearly all the bias current should be used in the output stage. The optimum value of α for relatively large C_L is approximately given by

$$\alpha_{opt} = \frac{1}{2}\sqrt{2}\sqrt{\frac{C_{be1}}{C_L}} \tag{4-28}$$

Using this optimal value of α, the maximum bandwidth can be approximated by

$$\omega_{n,opt} = \frac{1}{2}\sqrt{2}\frac{I_{SUP}}{V_T}\frac{1}{\sqrt{C_L C_{be1}}}\sqrt{\frac{C_{be1}}{C_L}} \tag{4-29}$$

where the last factor represents the bandwidth reduction relative to parallel compensation. Therefore, at high capacitive loads, the Miller compensation is accompanied by large bandwidth reduction which is the price that has to be paid for the insensitivity to parameter variations and the smaller size of the compensation capacitor. There are some techniques to limit the bandwidth reduction which are discussed in the next section. In Fig. 4-14 the plots are shown for a CMOS Miller-compensated two-stage amplifier with transistors operating in strong inversion. Clearly, the optimum is much broader in CMOS than in bipolar technology.

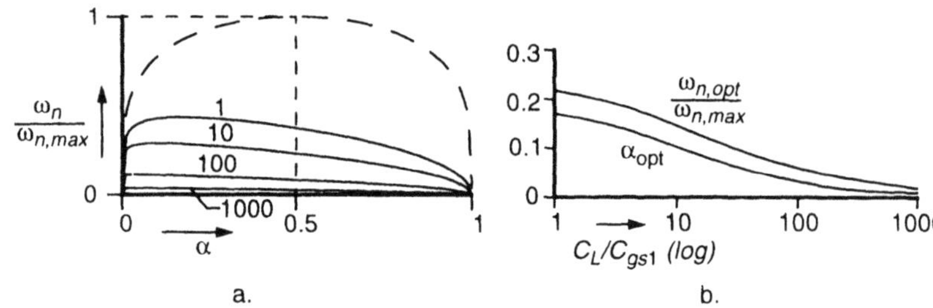

Fig. 4-14 *Bandwidth ω_n as a function of current division parameter α in CMOS technology of a Miller compensated two-stage amplifier for several values of C_L/C_{bel} (a) and the optimal α and resulting bandwidth $\omega_{n,opt}$ as a function of C_L/C_{bel} (b).*

Another disadvantage is the right-half plane zero that is caused by the direct signal transfer through the Miller capacitor. The frequency of the zero is given by

$$\omega_z = \frac{g_{m1}}{C_M} \tag{4-30}$$

which can also be derived by inspection when we assume that the zero is at that frequency where the direct transfer via the Miller capacitor is equal to the collector current of Q_1. Because the output voltage is negligible at very high frequencies, the voltage across C_M is nearly equal to the base-emitter voltage which immediately gives equation (4-30). The zero is in the right-half plane and is unwanted because it reduces the phase margin. However, before taking measures, we should first find out how serious the influence of the zero is. The influence on the phase margin depends on the frequency of the zero relative to the unity-gain frequency and can be expressed in degrees by [10]

$$\Delta\varphi = \mathrm{atan}\left(\frac{\omega_0}{\omega_z}\right) = \mathrm{atan}\left(\frac{1}{2}\frac{C_M}{C_L}\right) \tag{4-31}$$

which means that the influence of the zero is only important for relatively small load capacitors. If, for example, C_M is 1 *pF* and C_L is 5 *pF*, the reduction of the phase margin is only 6°. The simplest way to reduce the influence of the zero is to insert a resistor R_M

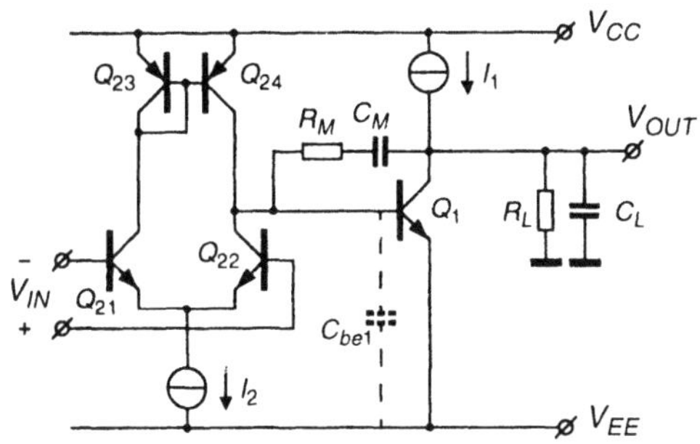

Fig. 4-15 *Two-stage amplifier in bipolar technology with Miller compensation and series resistor* R_M.

in series with the Miller capacitor, as shown in Fig.4-15. The resistor causes a zero in the left-half plane which counteracts the effect of the right-half-plane zero, yielding a combined zero at frequency

$$\omega_z = -\frac{\frac{1}{R_M} - g_{m1}}{C_M} \tag{4-32}$$

If we take R_M equal to $^1/g_{m1}$ the zero vanishes. However, if we apply this technique in a class-AB output stage, the zero will appear again in the left-hand-half plane if the output current increases. Therefore, another technique has been found to remove the zero which is discussed in section 4.5.3. Fortunately, the resistive technique can be applied in class-AB output stages as long as we make sure that the frequency $^1/R_MC_M$ which is the position of the zero at high output currents, is well above the band of the amplifier. It may even be advantageous to position the zero in the quiescent state in the left-hand-half plane. A point of concern is that the insertion of R_M also causes an additional pole. The pole can be calculated by assuming that R_M is relatively small so that the pole is found at high frequencies. In that case capacitors C_M and C_L can be regarded as short circuits yielding the third pole as given by

$$\omega_3 = -\frac{1}{R_M C_{be1}} \tag{4-33}$$

If R_M is equal to $1/g_{m1}$, the pole-zero spacing is C_M/C_{be1} which is usually sufficient to be sure that the influence of the vanishing of the zero is much larger than the influence of the third pole.

Finally, we discuss the behavior of a Miller-compensated amplifier under special conditions. For special conditions such as low resistive loads and heavy capacitive loads, the above-discussed equations are not accurate enough. In such a case, the complete characteristic equation as given by

$$\begin{aligned} P(s) &= 1 + s\left[r_{be1}C_{be1} + R_L C_L + (\beta_{F1}R_L + r_{be1})\,C_M\right] + \\ &\quad s^2 r_{be1} R_L \left(C_{be1}C_L + C_{be1}C_M + C_L C_M\right) \end{aligned} \tag{4-34}$$

should be the starting point for calculating the unity-gain frequency and the second pole. The only assumption we apply is that after introducing the Miller capacitor the poles are widely separated which is the purpose of the Miller capacitor. In that case, the first order term completely determines the dominant pole [11] and the unity-gain frequency is given by

$$\omega_0 = \frac{g_{m2}}{C_M\left(1 + \frac{1}{g_{m1}R_L}\right) + \frac{C_L}{\beta_{F1}} + \frac{C_{be1}}{g_{m1}R_L}} \tag{4-35}$$

At low resistive loads, the unity-gain frequency is reduced because the voltage gain $g_{m1}R_L$ of the output transistor is low. An advantage is that a smaller Miller capacitor is required. However, the accuracy of the unity-gain frequency is not so good because it partly depends on g_{m1} and R_L, and even C_{be1} may have influence. A more serious problem is that if the resistive load increases, the phase margin reduces because the unity-gain frequency increases. Therefore, g_{m1} should be adapted so that sufficient voltage gain $g_{m1}R_L$ is obtained. This may cost a lot of bias current especially if the load is as low as 50 ohms. Thus, in nearly all cases, the unity-gain frequency can accurately be expressed by

$$\omega_0 = \frac{g_{m2}}{C_M + \frac{C_L}{\beta_{F1}}} = \frac{g_{m2}}{C_M}\left(\frac{1}{1 + \frac{C_L}{\beta_{F1} C_M}}\right) \tag{4-36}$$

The unity-gain frequency is therefore well-controlled by g_{m2} and C_M. Only at very high capacitive loads the influence of C_L reduces the unity-gain frequency.

The second-order term is determined by the product of the dominant pole and the non-dominant pole. Therefore, the second pole can be calculated by dividing the second-order term by the first-order term. Using the above motivated condition that $g_{m1}R_L$ is large, we obtain

$$\omega_2 = -\frac{g_{m1}\left(1 + \frac{C_L}{\beta_{F1} C_M}\right)}{C_L + \frac{C_{be1}}{C_M} C_L + C_{be1}} \tag{4-37}$$

The second pole is only determined by g_{m1} and C_L for moderate values of C_L. If all the capacitors are of the same order, the frequency of the second pole is a factor three lower than the value obtained by equation (4-22). For very high values of C_L the second pole remains at a fixed position. It is interesting to consider the root locus for variations of C_L as shown in Fig. 4-16. For high values of C_L the dominant pole moves to the origin,

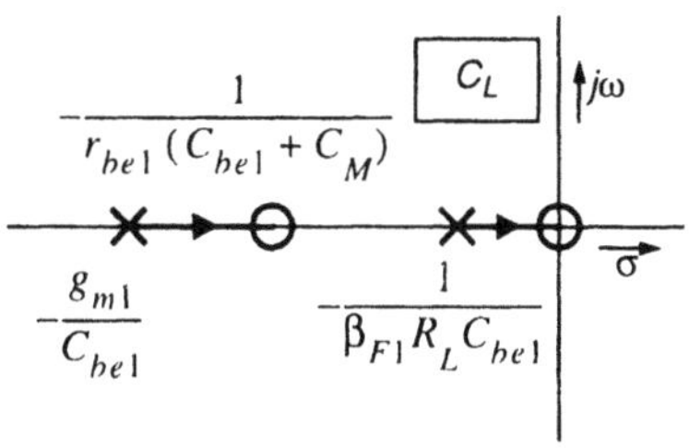

Fig. 4-16 *Root locus of Miller compensated amplifier for variations of C_L.*

while the second pole remains at a fixed position, as also can be concluded from the above equations. The result is that at a certain value of C_L the phase margin increases, while normally the phase margin decreases if C_L is increased. The value of C_L for which minimum phase margin is obtained is given by

$$C_L = \frac{C_M}{\beta_{F1}} \tag{4-38}$$

which can intuitively be understood by considering equation (4-36). The value given equation (4-38) is the transition point where the influence of C_L on the unity-gain frequency exceeds the influence of C_M. The minimum phase margin is given by

$$\varphi_{m,\,min} = 90 - \operatorname{atan}\left(\frac{\beta_{F1} g_{m2}}{4 g_{m1}}\right) \tag{4-39}$$

Thus, by selecting a proper ratio between g_{m2} and g_{m1}, sufficient phase margin can be obtained for any C_L independent of the total bias current used. Unfortunately, to obtain sufficient minimum phase margin the transconductance of the input stage should be almost a factor β_{F1} smaller than the transconductance of the input stage; this is very impractical to realize and severely limits the gain. The phase margin as a function of C_L for several values of the ratio g_{m1}/g_{m2} is plotted in Fig. 4-17. Although the biasing requirement is impractical, it is possible to ease the biasing requirement using cascoded Miller compensation which is discussed in the next section. Such an amplifier has been reported in CMOS technology [12].

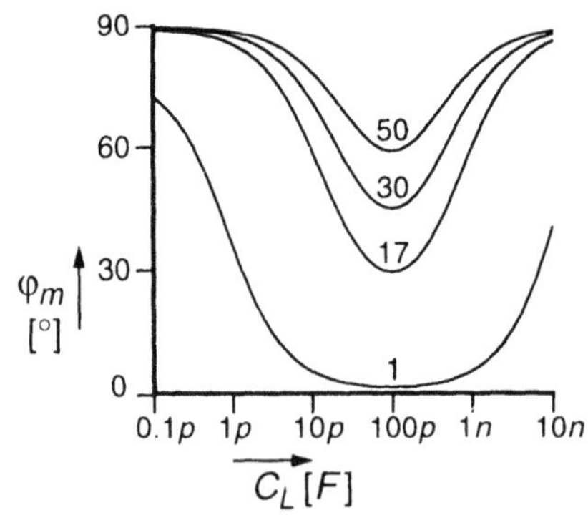

Fig. 4-17 *Phase margin φ_m as a function of capacitive load C_L for several values of the ratio g_{m1}/g_{m2}.*

4.5.3 Multipath-Miller-zero cancellation

Multipath-Miller-zero cancellation is a method to remove the right-half-plane zero of Miller compensation [13, 14]. This increases the phase margin and therefore the stability of a two-stage Miller compensated amplifier. The technique is applied in the circuit shown in Fig. 4-18. The amplifier is nearly identical to a conventional two-stage Miller-

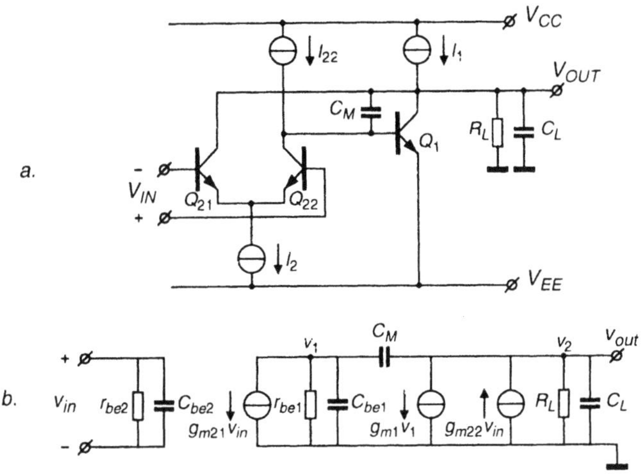

Fig. 4-18 *Circuit schematic of two-stage amplifier in bipolar technology with multipath-Miller-zero cancellation (a) and small-signal equivalent circuit (b).*

compensated amplifier consisting of input stage Q_{21}, Q_{22}, output stage Q_1, and Miller capacitor C_{M1}. However, only one output of the differential input stage is used to drive the output stage. The second output of the input stage is directly connected to the output node. In this way, the high-frequency feed-through of the input signal delivered by Q_{22} via the Miller capacitor is compensated by the signal delivered by Q_{21} to the output. This can also be concluded by calculating the numerator which is given by

$$N(s) = 1 + s\left(\frac{g_{m22}}{g_{m21}}\frac{C_{M1} + C_{be1}}{g_{m1}} - \frac{C_{M1}}{g_{m1}}\right) \qquad (4\text{-}40)$$

Thus, if g_{m21} and g_{m22} are equal, the zero is shifted to the frequency

$$\omega_z' = -\frac{g_{m1}}{C_{be1}} \tag{4-41}$$

which is situated in the left-half plane. Therefore, the technique only relies on the matching of transconductances and the cancellation is independent of the output-stage transconductance which is an important advantage over the resistive technique discussed in the previous section, if applied in class-AB output stages. However, the implementation of the feedforward path requires more bias current.

4.5.4 Active Miller compensation techniques

The disadvantages of Miller compensation as discussed in the previous section 4.5.2 are first, the right-half-plane zero which reduces the phase margin, and, second, the reduction of the bandwidth. These problems are caused by the connection of the Miller capacitor across the base collector junction. It causes a direct transfer from base to collector which results in the zero and, further, the Miller capacitor loads the output node and the base causing a bandwidth reduction. Both effects can be eliminated by using a voltage follower between the output node and the Miller capacitor, and a current follower between the Miller capacitor and the base of the output transistor. Practically, the voltage follower is difficult to implement without reducing the output-voltage range of a rail-to-rail output stage. Fortunately, the voltage follower is not necessary since the loading of the output node by the Miller capacitor is usually negligible compared to the load capacitor. The current follower, on the other hand, can effectively be implemented by a cascode which is often already present. An example of a two-stage amplifier with cascoded Miller compensation is depicted in Fig. 4-19 [15, 16].In fact, a nested-Miller compensation structure is applied, as will be explained later. For the time we assume C_{M2} is equal to the Miller capacitor C_M, and C_{M1} is zero. If the cascode behaves as an ideal current buffer the transfer function is given by

$$\frac{v_{out}}{v_{in}} = \frac{g_{m3}\beta_{F1}R_L}{1 + s(r_{be1}C_{be1} + R_LC_L + \beta_{F1}R_LC_M) + s^2 r_{be1}R_L(C_{be1}C_L + C_{be1}C_M)} \tag{4-42}$$

where g_{m3} is the transconductance of the input stage. As we see, the second order term is slightly increased compared to parallel compensation. However, if C_L is large, the bandwidth reduction is minimal. Unfortunately, the cascode adds another pole because its input impedance is not negligible. The main problem of the third pole is that if the transconductance of the output stage is increased, the third pole moves towards the

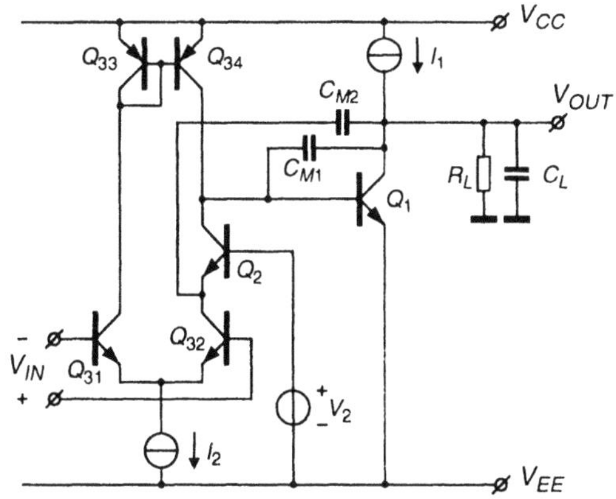

Fig. 4-19 *Two-stage amplifier in bipolar technology with nested-cascoded-Miller compensation.*

second pole and eventually complex poles are formed [16]. Therefore, a simple cascoded-Miller compensation cannot be used in class-AB amplifiers where the transconductance of the output stage largely varies. Controlling the imaginary part of the non-dominant poles is possible using cascoded nested Miller compensation [16], consisting of an inner Miller capacitor which is directly connected across the base-collector junction of output-stage transistor Q_1, while an outer Miller capacitor is connected to the cascode as shown in Fig. 4-19. The structure is comparable to nested Miller compensation which is discussed in the next section. However, the intermediate stage is now a cascode, while the intermediate stage of nested Miller compensation usually provides gain. The circuit is a mix of basic Miller compensation and cascoded Miller compensation and represents a trade-off between obtainable bandwidth and the maximum allowed imaginary parts of the non-dominant poles.

The simplified characteristic equation is given by

$$P(s) = 1 + s\beta_{F1}R_L(C_{M2} + C_{M1}) + s^3\frac{1}{g_{m2}}r_{be1}R_LC_{M1}C_{M2}C_L \qquad (4\text{-}43)$$
$$+ s^2\left[\frac{1}{g_{m2}}\beta_{F1}R_LC_{M1}C_{M2} + r_{be1}R_LC_L(C_{M1} + C_{be1}) + r_{be1}R_LC_{be1}(C_{M1} + C_{M2})\right]$$

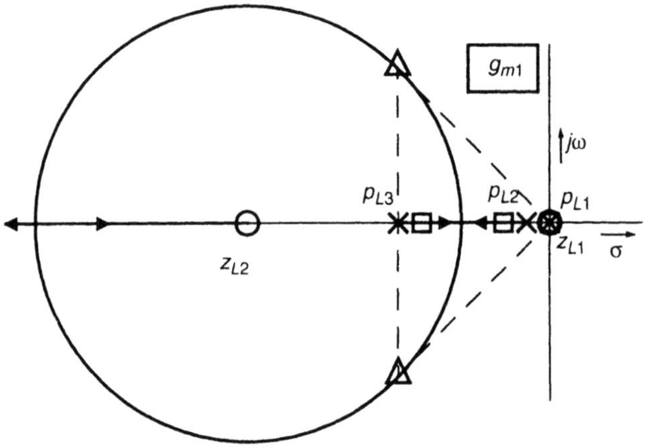

Fig. 4-20 *Root locus of a nested-cascoded-Miller compensated amplifier for variations of* g_{m1}.

The resulting root locus plot as shown in Fig. 4-20 contains a second zero at a relatively low frequency so that the imaginary part of the non-dominant poles is reduced. The zero is given by

$$z_{L2} = -\left(1 + \frac{C_{M1}}{C_{M2}}\right)\frac{g_{m2}}{C_{M1}} \tag{4-44}$$

The first pole and the first zero are situated in the origin. The second pole is given by

$$p_{L2} = -\frac{1}{R_L\left(C_{M1} + C_{M2} + C_L + \frac{C_{M1}}{C_{be1}}C_L\right)} \tag{4-45}$$

and is usually at a relatively low frequency compared to the third pole. The third pole of the locus is

$$p_{L3} = -\frac{g_{m1}}{C_{M1}}\frac{C_{be1}(C_{M1} + C_{M2}) + C_L(C_{M1} + C_{be1})}{C_{M2}(C_L + C_{be1})} \tag{4-46}$$

The maximum imaginary value of the poles is strongly limited by the zero z_{L2}. To maintain sufficient phase margin for all values of the output-stage bias current, the position of the poles with minimal damping must be the same as the open-loop positions of a third-order Butterworth polynomial, as given by Table 4-2. Thus, we conclude that the imaginary part of the poles must equal the real part at the minimal damping position. It can be shown that this is the case under the condition

$$z_{L2} = 2p_{L3} \tag{4-47}$$

The poles are then positioned as indicated by the squares in Fig. 4-20, with real part equal to the value of p_{L3}. Together with the dominant pole, the complex poles form a third-order open-loop Butterworth position if

$$|p_{L3}| = 2\omega_0 \tag{4-48}$$

The above conditions lead to the following design equations

$$C_{M1} = \frac{1}{2}(C_{M1} + C_{M2}) - C_{be1} \tag{4-49}$$

and

$$g_{m2} = 4\frac{C_{M1}}{C_{M1} + C_{M2}}g_{m3} \tag{4-50}$$

The value of C_{M1} is therefore slightly smaller than C_{M2}, and sufficient transconductance of the cascode is required. If the cascode is designed with a higher transconductance than given by equation (4-50), the value of C_{M1} can be smaller than the value calculated by equation (4-49). To complete the design equations we have to return to the quiescent state. In that case the second pole can be found as

$$p_2 = -\frac{2g_{m1}}{C_L} \tag{4-51}$$

To obtain second-order open-loop pole positions, the sum of the Miller capacitors should be

$$C_{M1} + C_{M2} = \frac{g_{m3}}{g_{m1}}C_L \tag{4-52}$$

Comparing the second pole position to the second pole of Miller compensation, we find that a factor of two higher bandwidth is obtained using nested-cascoded-Miller compensation.

Finally, the third pole should be calculated in the quiescent state to investigate its influence on the phase margin. The third pole is given by

$$p_3 = -\frac{1}{2}\frac{g_{m2}}{C_{M1}}\left(1 + \frac{C_{M1}}{C_{M2}}\right) \tag{4-53}$$

and should be well above the unity-gain frequency so that the phase margin is determined by the second pole. This is another reason why the transconductance of the cascode should be large.

4.6 Three-stage amplifier optimalization

In most cases, two stages are not sufficient to realize the required specifications of the amplifier. Especially in bipolar technology the gain of a two-stage amplifier is too small. Another example is the supply-voltage specification. The lowest possible supply voltage cannot be achieved using a two-stage design due to the cascode that drives the output stage in order to obtain sufficient gain. Thus, amplifiers operating at very low supply voltages require at least three stages. Without exception, the BiCMOS opamp designs have a three-stage topology. This is because the output-stage is always implemented in bipolar technology to take advantage of the high transconductance, while the low input impedance of the bipolar output transistor necessitates a CMOS intermediate stage to boost the gain.

The three-stage designs use the well-known nested-Miller compensation to stabilize the frequency response. Fortunately, nested-Miller compensation is a straight-forward extension of Miller compensation. Therefore, we can easily elaborate on the Miller theory of the previous section. This is done in section 4.6.1. Two new variations of nested-Miller compensation are active versions and can be compared to cascoded-Miller compensation. The first technique is called cascoded-nested-Miller compensation and is presented in section 4.6.2. Another possibility that is enabled by a three-stage topology is the application of a current mirror instead of a cascode. The resulting mirrored-nested-Miller compensation is discussed in section 4.6.2. Finally, in section 4.6.4 it is shown how multipath-nested-Miller compensation can be applied to combine the gain of a three-stage topology with the bandwidth of a two-stage circuit, while in section 4.6.5

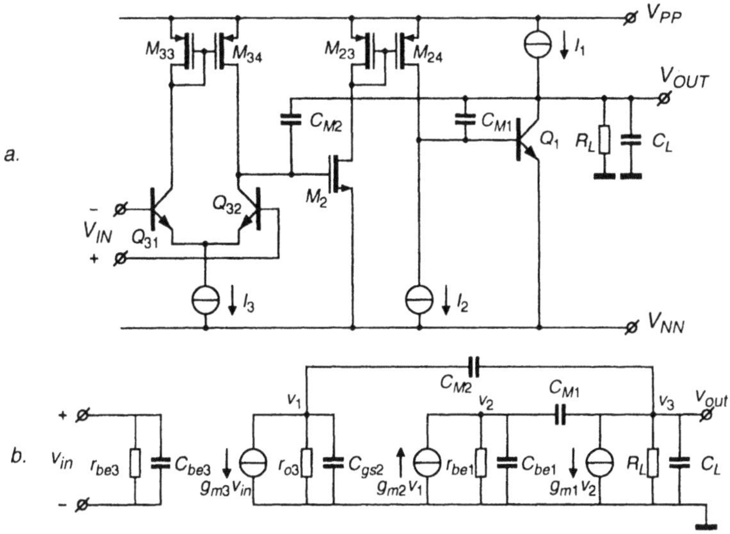

Fig. 4-21 *Circuit schematic of three-stage amplifier in BiCMOS technology with nested-Miller compensation (a) and small-signal equivalent circuit (b).*

other multipath compensation techniques are added to cancel the zeros caused by the Miller capacitors.

4.6.1 Nested-Miller compensation

As an example to demonstrate the three-stage compensation schemes, a BiCMOS amplifier is used with bipolar input stage, bipolar output stage, and a CMOS intermediate stage as depicted in Fig. 4-21a. A bipolar output stage is advantageous because a high bandwidth is obtained using the relatively high transconductance of the bipolar transistor. The high input impedance of the CMOS transistor can be exploited by applying a CMOS input stage where the gain can easily be increased by cascoding the input stage. Further, the bipolar input stage is applied to achieve low offset voltage and noise voltage. PMOS current mirrors are preferred over PNP current mirrors because of the absence of base-currents and because no degeneration resistors are required to reduce the contribution to offset and noise yielding a compact circuit. The small-signal equivalent circuit is shown in Fig. 4-21b. The nested-Miller compensation consists of a Miller capacitor C_{M1} across the output transistor Q_1, which is also found in the two-stage-Miller topology and a second Miller capacitor C_{M2} [17,18,8,7]. The two capacitors are required to shift the three poles found at the gate of M_2, the base of Q_1, and the output node, in such a way that a single dominant pole is obtained. The inner-Miller capacitor C_{M1} shifts the two poles around Q_1 apart, yielding a single dominant pole. Together with the pole at the gate of M_2, two dominant poles still exist in the three stage amplifier after insertion of C_{M1}. The two remaining poles can be split apart using the outer-Miller capacitor C_{M2}, resulting in the desired single-dominant-pole response. The unity-gain frequency can easily be found in the same way as with Miller compensation by assuming the high-frequency response is determined by the overall Miller capacitor C_{M2}. Thus, the input stage transforms the input signal into a current as determined by its transconductance g_{m3}. Due to the loop-gain around C_{M2} generated by M_2 and Q_1, all the signal from the input stage flows through C_{M2} and is transformed into the output voltage. The high-frequency transfer function is then given by

$$\frac{v_{out}}{v_{in}} = \frac{g_{m3}}{sC_{M2}} \tag{4-54}$$

which gives the unity-gain frequency

$$\omega_0 = \frac{g_{m3}}{C_{M2}} \tag{4-55}$$

Since there is now more gain than in a two-stage circuit, equation (4-55) accurately describes the unity-gain frequency. Only in extreme cases other capacitors such as C_L may have influence. The behavior of the non-dominant poles can be investigated by considering the circuit for high frequencies so that C_{M2} can be regarded as a short circuit. In that case, the combination of output stage and intermediate stage form a two-stage Miller compensated circuit with unity-gain feedback. The characteristic equation of the two-stage amplifier with unity-gain feedback is easily calculated from the open-loop equation, as given by equation (4-34), by dividing the first and second-order terms by the dc-gain, and removing the non-dominant terms. Further, the dominant pole is approximately situated in the origin, yielding the simplified characteristic equation of a nested-Miller compensated amplifier as given by

$$P(s) = sg_{m2}r_{o3}\beta_{F1}R_LC_{M2}\left(1 + s\frac{C_{M1}}{g_{m2}} + s^2\frac{C_{M1}C_L}{g_{m1}g_{m2}}\right) \tag{4-56}$$

Using the results listed in Table 4-2 and Table 4-3, the characteristic equation should be equal to

$$P(s) = s\frac{A_0}{\omega_0}\left(1 + s\frac{1}{2\omega_0} + s^2\frac{1}{8\omega_0^2}\right) \qquad (4\text{-}57)$$

The resulting values for the Miller capacitors can be expressed as a function of the load capacitor by

$$C_{M1} = 2\frac{g_{m2}}{g_{m1}}C_L \qquad (4\text{-}58)$$

and

$$C_{M2} = 4\frac{g_{m3}}{g_{m1}}C_L \qquad (4\text{-}59)$$

The unity-gain frequency as a function of the load capacitor is given by

$$\omega_0 = \frac{1}{4}\frac{g_{m1}}{C_L} \qquad (4\text{-}60)$$

so that the obtainable bandwidth is approximately a factor two smaller than the bandwidth of a two-stage Miller-compensated amplifier. If we calculate the characteristic equation of the nested-Miller compensated amplifier, a more accurate result is obtained. A simplified version is given by

$$P(s) = 1 + sg_{m2}r_{o3}\beta_{F1}R_LC_{M2} + s^2(g_{m1} - g_{m2})r_{be1}r_{o3}R_LC_{M1}C_{M2} + s^3r_{be1}r_{o3}R_LC_LC_{M2}(C_{M1} + C_{be1}) \qquad (4\text{-}61)$$

The factor g_{m1}-g_{m2} in the second-order term indicates that the transconductance of the output stage should be much larger than the transconductance of the intermediate stage. In that case the second-order term is equal to the term in equation (4-56). The bandwidth of the closed-loop unity-gain circuit ω_n can be calculated from the third-order term yielding

$$\omega_n = \sqrt[3]{\frac{g_{m1}g_{m2}g_{m3}}{C_L C_{M2}(C_{M1}+C_{be1})}} \tag{4-62}$$

Using equation (4-55) and noting that the unity-gain frequency should be equal to half ω_n, we obtain

$$\omega_n = \frac{1}{2}\sqrt{2}\sqrt{\frac{g_{m1}g_{m2}}{C_L(C_{M1}+C_{be1})}} \tag{4-63}$$

which is nearly identical to equation (4-26) of Miller compensation, except for the factor $^1/_2\sqrt{2}$. Therefore, the biasing considerations for maximum bandwidth are the same as with Miller compensation. However, the equations in BiCMOS technology are complicated and it is not possible to calculate accurately a general result if the intermediate stage operates in strong inversion. It is possible to estimate how much current should be biased in the intermediate stage. The transconductance g_{m2} is small and therefore C_{M1} is small. This means that the optimum is found close to the optimum of a parallel compensated circuit. For a mixed BiCMOS circuit, this optimum is found for $\alpha = {}^1/_3$ which means that $^1/_3$ of the bias current should be used in the intermediate stage. In most cases, C_{M1} is not completely negligible compared to C_{be1} and therefore the optimum is at a somewhat smaller value.

Further, the zeros may have important influence. The zeros are determined by the numerator polynomial of the transfer function as given by

$$N(s) = 1 - s\left(\frac{C_{M2}}{g_{m2}\beta_{F1}} + \frac{C_{M1}}{g_{m1}}\right) - s^2\frac{C_{M1}C_{M2}}{g_{m1}g_{m2}} \tag{4-64}$$

which can be rapidly found by inspection. At high frequencies the output voltage is negligible so that the voltage across C_{M2} is equal to the gate-source voltage of M_2, and the voltage across Q_1 is equal to the base-emitter voltage of Q_1. The value of the zeros results from the condition that the signal current flowing through capacitors C_{M1} and

C_{M2} is equal to the collector current of Q_1 yielding equation (4-64). If C_{M1} is not too small, the zeros are approximately given by

$$z_1 = \frac{g_{m1}}{C_{M1}} \quad \text{and} \quad z_2 = -\frac{g_{m2}}{C_{M2}} \tag{4-65}$$

The first zero is situated in the right-half plane at the same position as with Miller compensation. A second zero exists in the left half-plane at a lower frequency. This zero improves the phase margin, but care must be taken that the zero is placed outside the band of the amplifier.

4.6.2 Cascoded-nested-Miller compensation

In the same way as with Miller compensation, a cascode may be inserted in the loop created by C_{M2} yielding cascoded-nested-Miller compensation. The circuit is shown in Fig. 4-22. A bipolar cascode is applied because its input impedance is smaller so that the frequency of parasitic pole due to the loading of the Miller capacitor is high. We might expect a bandwidth improvement, but later we find that, at least with the standard version, no bandwidth improvement is obtained. The remaining advantage of cascoded-nested-Miller compensation is the improvement of high-frequency distortion as is discussed in chapter 6. Another reason for discussing cascoded-nested-Miller compensation is that it is the first step towards the mirrored-nested-Miller compensation structure which is discussed in the next section.

With the cascode inserted in the outer Miller loop, a dominant pole is still obtained using C_{M2} in the same way as with conventional nested-Miller compensation. However, because the cascode isolates the gate of M_{21} from C_{M2}, the Miller capacitor C_{M2} does not function as a unity-gain feedback for high frequencies around the internal two-stage amplifier Q_1, M_{21},M_{22} which is compensated using C_{M1}. In fact, gain is created in the loop of C_{M2} for which the internal two-stage amplifier is normally not compensated, yielding complex poles. The output voltage is transformed into a current by C_{M2}. The current flows through the cascode and is transformed into a voltage again by C_{P2}. Capacitor C_{P2} can consist of the parasitic gate-source capacitors of the intermediate stage, or can be a capacitor that is deliberately inserted to reduce the high-frequency gain. Now, the high-frequency gain is determined by the ratio C_{P2} over C_{M2}. The characteristic equation of cascoded-nested-Miller compensation can be easily derived from equation (4-56) by introducing the gain around the internal two-stage amplifier.

$$P(s) = s g_{m2} r_{o3} \beta_{F1} R_L C_{M2}\left(1 + s\frac{C_{P2}C_{M1}}{C_{M2}g_{m2}} + s^2\frac{C_{P2}C_{M1}C_L}{C_{M2}g_{m1}g_{m2}}\right) \tag{4-66}$$

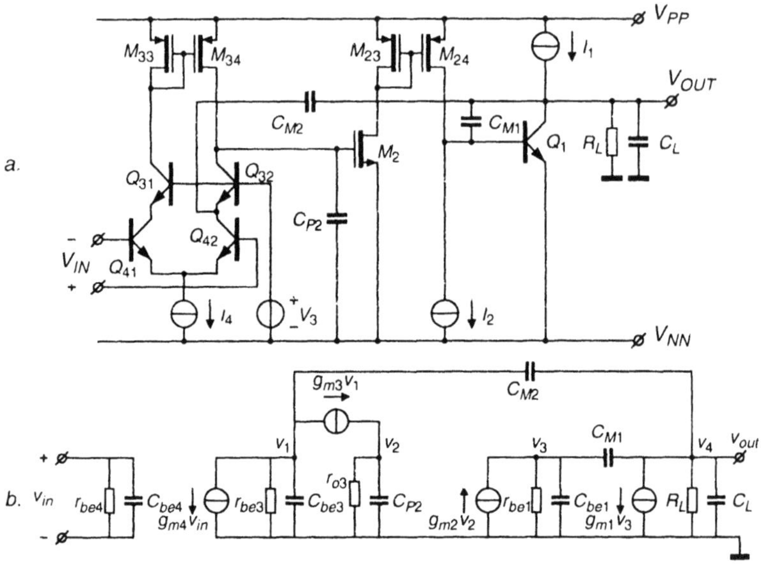

Fig. 4-22 *Circuit schematic of three-stage amplifier in BiCMOS technology with cascoded-nested-Miller compensation (a) and small-signal equivalent circuit (b)*

where r_{o3} is the output resistance of the cascoded input stage. By taking C_{P2} equal to C_{M2}, the equation reduces to the equation of conventional nested-Miller compensation. Thus, obtainable bandwidth and the bias current considerations are the same. The dimensioning of compensation capacitors is only identical to nested-Miller compensation if C_{M2} and C_{P2} are equal. However, different values of C_{P2} can be used and yield the same bandwidth. But, if C_{P2} is taken much too small, a very large value of C_{M1} is required. It is easy to show that there is an optimal value of C_{P2} for which the total capacitance is the smallest. The optimal C_{P2} value is given by

$$C_{P2} = \sqrt{\frac{1}{2}\frac{g_{m2}}{g_{m3}}}C_{M2} \tag{4-67}$$

Finally, the insertion of the cascode results in an additional node and therefore a high-frequency pole exists. This pole should be situated well outside the band of the amplifier and, therefore, it is necessary to calculate the pole frequency. Assuming the pole is situated well outside the frequency band, it is approximately given by

$$\omega_{p4} = -\frac{g_{m3}}{C_{M2}} \tag{4-68}$$

This means that the transconductance of the cascode should be considerably larger than the transconductance of the input stage. This is usually not the case, so that equation (4-68) is not valid. It is therefore better to consider the amplifier as a four-stage circuit. Then, the unity-gain frequency is given by

$$\omega_0 = \frac{1}{3.41}\frac{g_{m1}}{C_L} \tag{4-69}$$

and the capacitors are found as

$$C_{M1} = 1.71\frac{g_{m2}}{g_{m1}}\frac{C_{M2}}{C_{P2}}C_L \text{ and } C_{M2} = 3.41\frac{g_{m4}}{g_{m1}}C_L \tag{4-70}$$

A final finding is a relation between the transconductance of M_3 and the input-stage transconductance

$$g_{m3} = 6.83 g_{m4} \tag{4-71}$$

As in the case of two-stage nested-cascoded Miller compensation, we find that the transconductance of the cascode should be sufficiently high.

The advantage of cascoded-nested-Miller compensation is not a bandwidth improvement; however, due to the cascode, the Miller capacitor C_{M2} is isolated from the gate of M_2 which carries a distorted signal, because M_2 delivers the nonlinear base current of Q_1. Without the cascode this signal is directly transferred to the output by C_{M2} for higher frequencies. Thus, using cascoded-nested-Miller compensation a better mid-frequency distortion is obtained as discussed in section 4.7. Further, a bandwidth improvement may be obtained by inserting a resistor in series with C_{P2} to block the loading of the intermediate stage by C_{P2} for very high frequencies.

4.6.3 Mirrored-nested-Miller compensation

Instead of a cascode, also a current mirror can be inserted in the outer Miller loop of nested-Miller coinpensation. The current mirror replaces the inverting action of the current mirror M_{21} M_{22} of nested Miller compensation, as shown in Fig. 4-21. The result is shown in Fig. 4-23 and is called mirrored-nested-Miller compensation where the current mirror consists of transistors Q_{31}, Q_{32}. A bipolar current mirror is used so that

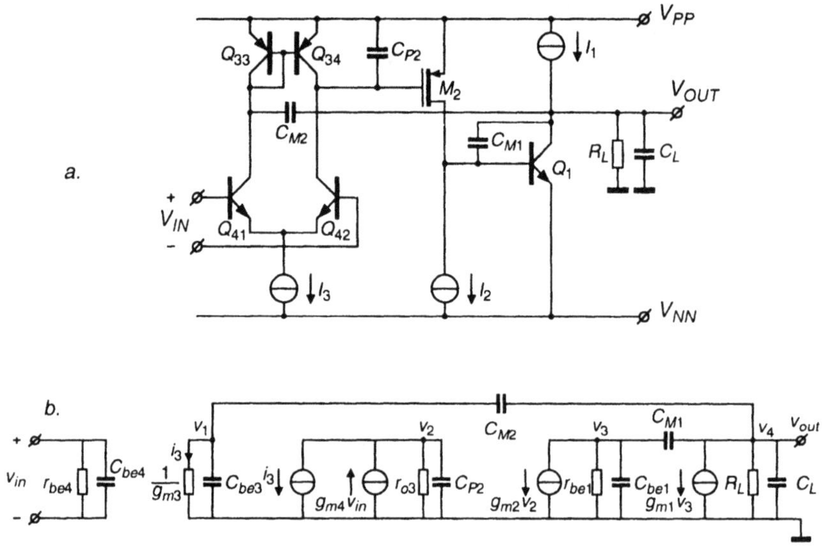

Fig. 4-23 *Circuit schematic of three-stage amplifier in BiCMOS technology with mirrored-nested-Miller compensation (a) and small-signal equivalent circuit (b).*

the C_{M2} is connected to a low impedance. The current mirror is usually available as part of a summing circuit that sums the differential signals of the input stage. Therefore, the replacement of the current mirror between input stage and intermediate stage of nested-Miller compensation by a current mirror in the outer Miller loop represents a simplification of three-stage amplifier design. Since the topology is identical to cascoded-nested-Miller compensation, all the equations from Eq. 4-66 to Eq. 4-68 are applicable to mirrored-nested-Miller compensation.

It is interesting to compare nested-Miller compensation and mirrored-nested-Miller compensation with respect to their only difference, the position of the parasitic fourth pole. In the mirrored-nested-Miller circuit, the fourth pole is given by equation (4-68).

The parasitic pole of nested-Miller compensation is caused by the current mirror connected between output stage and intermediate stage. Since the mirror is not loaded by the compensation capacitors, the pole is usually situated at a sufficiently high frequency and approximately given by

$$p_4 = -\frac{g_{m23}}{C_{gs23} + C_{gs24}} \tag{4-72}$$

where g_{m23} is the transconductance of M_{23} which is shown in Fig. 4-21, and C_{gs23} and C_{gs24} are the gate-source capacitances of M_{23} and M_{24}, respectively. Therefore, the simplification of the circuit into a mirrored-nested-Miller topology normally results in a deterioration of the frequency response. Especially in amplifiers designed for small noise voltage, where the transconductance of the current mirror in the summing circuit is much smaller than the transconductance of the input stage, problems may be expected. On the other hand, if nested-Miller compensation is used in an amplifier with bipolar output stage, the current-mirror transistors are usually quite large in order to handle the large base currents of the output transistors. In the quiescent state the current mirror transistors are biased by a relatively small bias current so that their transit frequency is small. In that case, the mirrored-nested-Miller compensation achieves comparable, or even improved performance.

4.6.4 Multipath-Nested-Miller compensation

The multipath technique can also be used to cancel a pole. In Fig. 4-24 the multipath technique is applied to a three-stage nested-Miller amplifier [19,20,21,7]. The circuit consists of a normal nested-Miller compensated amplifier consisting of input stage Q_{31},Q_{32}, M_{33}, M_{34}, intermediate stage M_2, output stage Q_1, and frequency compensation C_{M1} and C_{M2}. The multipath is created using a second input stage Q_{41},Q_{42}. Thus, an independent feedforward path is created that directly drives the output stage. In fact, a three-stage circuit and a two-stage circuit are now connected in parallel, combining the advantages of both topologies, high gain and high bandwidth, respectively. We now consider the frequency response of the feedforward path and the complete amplifier. Again, the addition of the feedforward input stage does not alter the denominator of the circuit but only the numerator. Thus, the denominator is identical to the characteristic equation of nested-Miller compensation. The frequency response of the nested-Miller compensated amplifier is indicated in Fig. 4-25 by the solid line. The feedforward path is connected to the same circuit and therefore has the same characteristic equation and the same poles ω_{p1}-ω_{p3}. The bypass of the intermediate stage creates a zero ω_z. If the zero is positioned correctly, the feedforward path takes over at the frequency of the second pole ω_{p2}, as shown in Fig. 4-25 by the dashed line, yielding a smooth frequency response

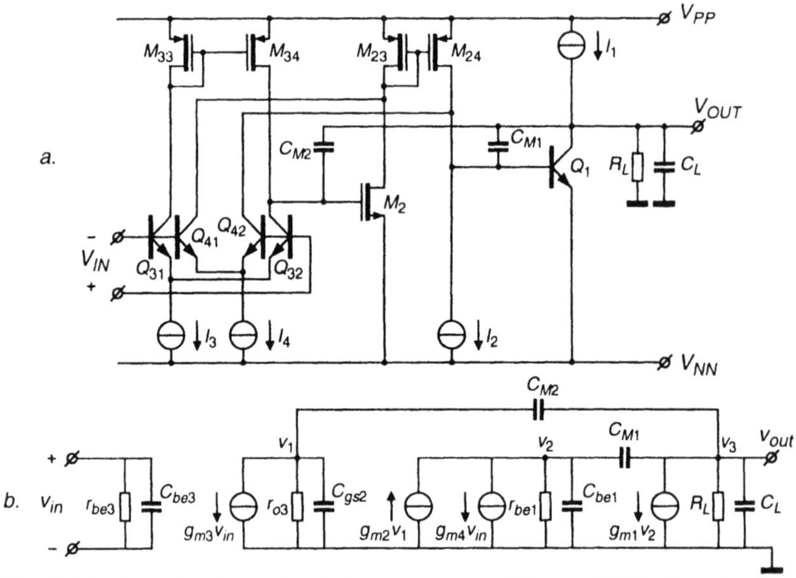

Fig. 4-24 *Circuit schematic of three-stage amplifier in BiCMOS technology with multipath-nested-Miller compensation (a) and small-signal equivalent circuit (b).*

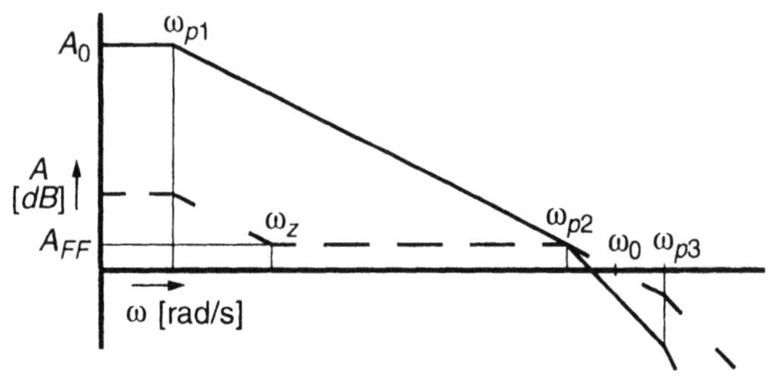

Fig. 4-25 *Frequency response of three-stage amplifier with multipath-nested-Miller compensation. Solid line indicates characteristic of circuit without feedforward path and dashed line indicates characteristic of feedforward path.*

with only two poles, ω_{p1} and ω_{p3}. The two characteristics intersect at the second pole if the gain-bandwidth product of the low-frequency path is equal to the unity-gain frequency of the high-frequency feedforward path. Both parameters are determined by the transconductances of the two input stages and the values of the two Miller capacitors, yielding the requirement

$$\frac{g_{m3}}{C_{M2}} = \frac{g_{m4}}{C_{M1}} \tag{4-73}$$

The requirement can be satisfied with good accuracy since transconductances can be well matched and also, capacitor values can be well matched. The above requirement is not entirely accurate. The gain-bandwidth of the low-frequency path is accurately determined by g_{m3} and C_{M2}, but the unity-gain frequency of the feedforward path is influenced by the feedback loop consisting of C_{M2}, M_2, and Q_1. This feedback loop determines the value of the mid-frequency magnitude of the feedforward path A_{FF} which is an important parameter that affects the maximum obtainable bandwidth. If A_{FF} is much smaller than $0dB$, the two high-frequency poles are complex and it is not possible to cancel the second pole. In that case, the bandwidth enhancement of the feedforward path is small so that the bandwidth is about the same as the bandwidth of a standard nested-Miller amplifier. When A_{FF} is large, the high-frequency response is dominated by the feedforward path, yielding a bandwidth that is comparable to a two-stage Miller compensated circuit. Thus, we first calculate A_{FF} and thereafter we investigate its influence on the bandwidth. The mid-frequency magnitude of the feedforward path A_{FF} can be found in an intuitive way. In the mid-frequency band the impedance at node v_1 is dominated by the capacitors C_{M2} and C_{gs2}, while the internal amplifier Q_1 is still operating with high gain. In this frequency range the feedforward-path can be simplified as shown in Fig. 4-26.

Due to the gain of the feedback loop consisting of Q_1, C_{M2}, and M_2, the signal current of $g_{m4}v_{in}$ is regulated equal to the signal current $g_{m4}v_1$ of transistor M_2. The output voltage is attenuated by the capacitive division of C_{M2} and C_{gs2} which determines the signal voltage at the gate. Combining the results gives

$$A_{FF} = \frac{g_{m4}}{g_{m2}} \frac{(C_{gs2} + C_{M2})}{C_{M2}} \approx \frac{g_{m4}}{g_{m2}} \tag{4-74}$$

It is possible to derive a requirement in order to prevent complex poles. The poles are becoming complex at the point where ω_{p2} and ω_{p3} have the same frequency. In order to obtain a phase margin of the order of 60°, the amplifier is usually designed in such a way that the second pole (in this case ω_{p3}) frequency is a factor of two larger than the unity-

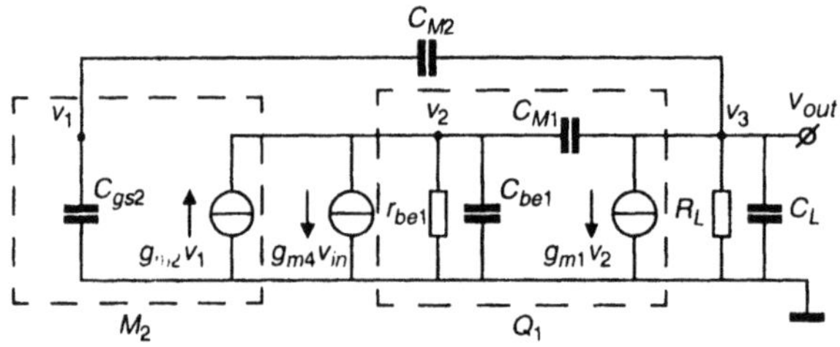

Fig. 4-26 *Simplified small-signal equivalent circuit of the feedforward path of multipath-nested-Miller compensation.*

gain frequency. Therefore, the magnitude of the feedforward path A_{FF} should be larger than $^1/_2$ yielding the requirement

$$\frac{g_{m4}}{g_{m2}} > \frac{1}{2} \tag{4-75}$$

The requirement states that the transconductance of the intermediate stage should be smaller than the transconductance of the input stage.

A more accurate value of the unity-gain frequency can be found by considering the influence of the feedback through C_{M2} on the unity-gain frequency of the internal Miller-compensated output stage using Fig. 4-26. Assuming the high-frequency response of the output stage is determined by C_{M1}, the transfer function of the feedforward path can be estimated by only considering the output stage as an integrator, with integration capacitor C_{M1} and input currents $g_{m4}v_{in}$ and $g_{m2}v_1$ yielding

$$v_{out} = sC_{M1}\left(g_{m4}v_{in} - g_{m2}\frac{C_{M2}}{C_{M2} + C_{gs2}}v_{out}\right) \tag{4-76}$$

Using equation (4-76) the unity-gain frequency is calculated

$$\omega_0 = \frac{g_{m4}}{C_{M1}}\sqrt{1 - \left(\frac{g_{m2}}{g_{m4}}\frac{C_{M2}}{C_{M2} + C_{gs2}}\right)^2} = \frac{g_{m4}}{C_{M1}}\sqrt{1 - \frac{1}{A_{FF}^2}} \tag{4-77}$$

Clearly, parameter A_{FF} reduces the unity-gain frequency of the amplifier. This means that for the same unity-gain frequency a larger capacitor C_{M1} is required which reduces the obtainable bandwidth ω_n. The bandwidth ω_n can be calculated by dividing the third-order term of the nested-Miller circuit from equation (4-61) by the value of the second pole and the loop gain. The second pole can accurately be calculated by finding the intersection of A_{FF} and the response of the low-frequency path yielding

$$p_2 = -\frac{g_{m3}}{g_{m4}} \frac{g_{m3}}{(C_{gs2} + C_{M2})} = -\frac{\omega_0}{A_{FF}} \tag{4-78}$$

Then, the bandwidth is found as

$$\omega_n = \sqrt{\frac{g_{m1} g_{m3}}{C_L (C_{be1} + C_{M1})}} \tag{4-79}$$

which is nearly identical to the bandwidth of a two-stage Miller compensated amplifier. However, note that according to equation (4-26), the value of the Miller capacitor is larger resulting in a smaller bandwidth. If, for example, the value of A_{FF} is 2 the bandwidth reduction is only 7%. Increasing the value of A_{FF} lowers the frequency of the second pole which is also the frequency of the pole-zero doublet. Although the pole-zero spacing is relatively small due to the good matching, a doublet still remains which increases the settling time of the amplifier. The pole-zero doublet introduces a slow-settling component in the step response with settling time [22,7]

$$T_S = \frac{\ln\left(\frac{S_d \omega_d}{a \omega_0}\right)}{\omega_d} \tag{4-80}$$

where S_d is the relative pole-zero spacing of the doublet, a is the required settling accuracy, and ω_d is the frequency of the doublet. Introducing the value of the doublet which equals ω_{p2} gives

$$T_S = A_{FF} \frac{\ln\left(\frac{S_d}{a A_{FF}}\right)}{\omega_0} \tag{4-81}$$

As we see, the settling time reduces if the value of A_{FF} is lowered. The influence of the pole-zero doublet of multipath compensation is further discussed in section 4.8.

The multipath technique can be applied in any other amplifier with a nested-Miller topology. As an example we show the application of the multipath technique in a mirrored-nested-Miller amplifier. The multipath-mirrored-nested-Miller circuit is depicted in Fig. 4-27. The topology is largely comparable to the multipath-nested-Miller

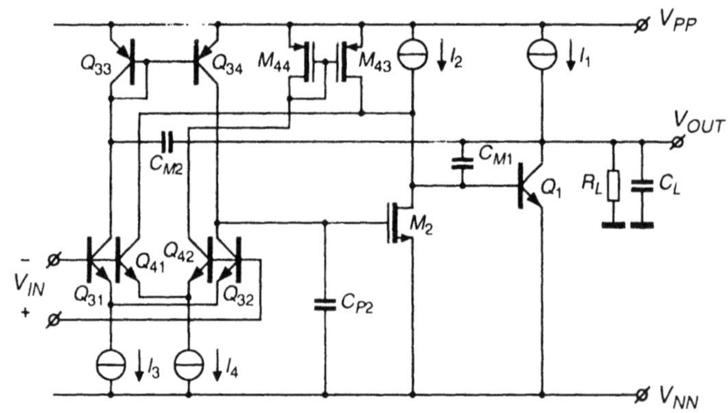

Fig. 4-27 *Circuit schematic of three-stage amplifier in BiCMOS technology with multipath-mirrored-nested-Miller compensation.*

circuit yielding the same equations. Only the parameter A_{FF} is different due to the mirror which is inserted in the loop containing C_{M2}. In the mirrored-nested-Miller circuit the output voltage is transformed into a current by C_{M2}. The current flows through current mirror Q_{33}, Q_{34} and is transformed into a voltage by C_{P2}. Therefore, the ratio C_{P2} over C_{M2} is present in the expression for A_{FF} which is given by

$$A_{FF} = \frac{g_{m4}}{g_{m2}} \frac{C_{P2}}{C_{M2}} \tag{4-82}$$

The parameter A_{FF} can therefore be manipulated by changing C_{P2} as well as g_{m2}. All the other equations are equal to the results found for multipath-nested-Miller compensation.

4.6.5 Multiple-multipath nested-Miller compensation

The ultimate multipath compensation structure would be the combination of multipath compensation with multipath miller-zero-cancellation. This idea is demonstrated in the three-stage BiCMOS amplifier topology as shown in Fig. 4-28 [14,23]. In the amplifier, transistor Q_{41} implements the multipath of multipath-nested-Miller compensation while

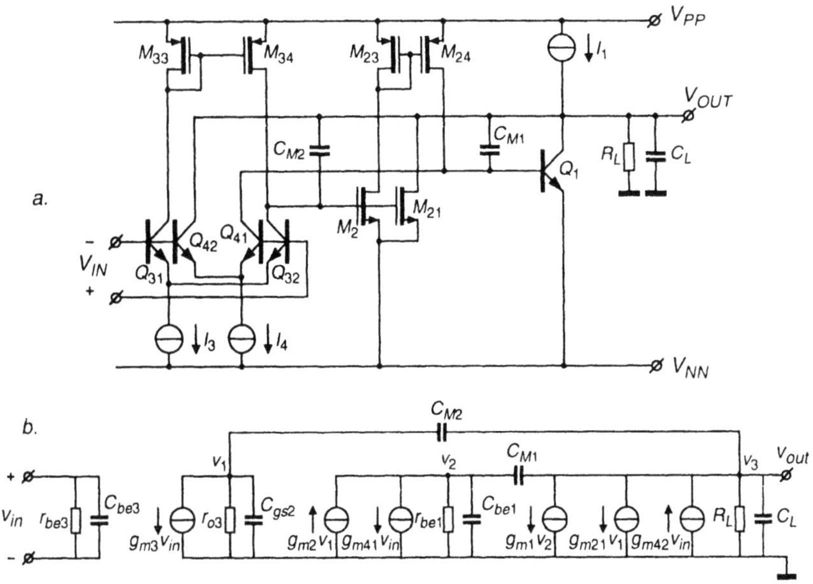

Fig. 4-28 *Circuit schematic of three-stage amplifier in BiCMOS technology with multipath-nested-Miller compensation (a) and small-signal equivalent circuit (b).*

transistors M_{21} and Q_{42} create miller-zero-cancellation feedforward paths. Transistor M_{21} cancels the high-frequency signal feed-through of M_2 via C_{M2}, while Q_{42} cancels the feed-through via C_{M2}, caused by the input stage Q_{31}, Q_{32} and the multipath stage Q_{41}. We briefly review the influence of each feedforward path on the transfer function of the amplifier. First, the influence of M_{21} is investigated. Since M_{21} does not create a multipath controlled by the input voltage but is controlled by an internal node, the addition of M_{21} does affect the characteristic equation of the circuit. The denominator can be calculated as

$$P(s) = 1 + s g_{m2} r_{o3} \beta_{F1} R_L C_{M2} + s^2 (g_{m1} - g_{m2} + g_{m21}) r_{be1} r_{o3} R_L C_{M1} C_{M2} + s^3 r_{be1} r_{o3} R_L C_L C_{M2} (C_{M1} + C_{be1}) \quad (4\text{-}83)$$

Thus, if g_{m21} is equal to g_{m2}, the second-order term of the characteristic equation slightly improves because the term containing $-g_{m2}$ is cancelled. This term is, however, relatively small compared to g_{m1}. The simplified numerator polynomial containing the influence of M_{22} is given by

$$N(s) = 1 + s\left[\frac{g_{m21}}{g_{m2}}\frac{C_{M1} + C_{be1}}{g_{m1}} - \frac{C_{M1}}{g_{m1}}\right] - s^2\frac{(C_{M1} + C_{be1})\,C_{M2}}{g_{m1}g_{m2}} \tag{4-84}$$

If g_{m21} is equal to g_{m2}, the second-order term of the numerator is made positive yielding a zero at high frequencies

$$\omega_{z1} = -\frac{g_{m1}}{C_{be1}} \tag{4-85}$$

However, since the second-order term is not changed, now the other zero is in the right-half plane

$$\omega_{z2} = \frac{g_{m2}}{C_{M2}}\frac{C_{be1}}{(C_{M1} + C_{be1})} \tag{4-86}$$

and is shifted to a rather low frequency. Therefore, the addition of M_{21} without other compensation techniques is not beneficial at all.

Next, a feedforward path can be introduced that drives the output stage, as discussed in the previous section. Addition of feedforward path yields the multipath-nested-Miller topology. The numerator of the multipath structure is given by

$$N(s) = 1 + s\left[r_{o3}(C_{M2} + C_{gs2}) - \frac{g_{m1}}{C_{M1}}\right] - s^2\frac{r_{o3}(C_{M2} + C_{gs2})\,C_{M1}}{g_{m1}} \tag{4-87}$$

The feedforward path to the base of Q_1 creates the multipath zero required to cancel one of the poles of the transfer function. Further, a right-half plane zero is present which is approximately at the same frequency as the right-half-plane zero of conventional Miller compensation. Combination of the two discussed multipath techniques does not make sense since they both affect the first-order term of the numerator, and the effect of the first technique is negligible compared to the second one.

The final multipath that can be created in the nested-Miller circuit directly drives the output node using transistor Q_{42}. Using this feedforward path the direct transfer via the Miller capacitors to the output can be compensated. The numerator is given by

$$N(s) = [1 + sr_{be1}(C_{M1} + C_{be1})]\,[1 + sr_{o3}(C_{M2} + C_{gs2})] \qquad (4\text{-}88)$$

and contains second-order terms that can cancel the negative second-order terms of the previously discussed numerator polynomials. The numerator containing all contributions of the three feedforward paths is obtained by taking into account the different values of the dc-gains of the transfer functions yielding the simplified numerator polynomial

$$N(s) = 1 + s\left[\frac{g_{m21}}{g_{m2}}\frac{C_{M1} + C_{be1}}{g_{m1}} - \frac{C_{M1}}{g_{m1}} + \frac{g_{m41}}{g_{m3}}\frac{C_{M2} + C_{gs2}}{g_{m2}}\right] + \qquad (4\text{-}89)$$
$$s^2\left[\frac{g_{m42}}{g_{m3}}\frac{(C_{M2} + C_{gs2})(C_{M1} + C_{be1})}{g_{m1}g_{m2}} - \frac{(C_{M1} + C_{be1})C_{M2}}{g_{m1}g_{m2}} - \frac{g_{m41}}{g_{m3}}\frac{(C_{gs2} + C_{M2})C_{M1}}{g_{m1}g_{m2}}\right]$$

There are now two possible combinations that are of interest. The first one is the combination of M_{22} with Q_{42} which can be considered as multipath-Miller-zero-cancelled nested-Miller compensation. If $g_{m21} = g_{m2}$ and $g_{m41} = g_{m3}$, the resulting simplified numerator is

$$N(s) = 1 + s\frac{C_{be1}}{g_{m1}} + s^2\frac{C_{gs2}(C_{M1} + C_{be1})}{g_{m1}g_{m2}} \qquad (4\text{-}90)$$

which describes two zeros that are situated in the left half-plane at relatively high frequencies. The other possibility is the combination of Q_{41} and Q_{42} which can be named multipath-Miller-zero-cancelled multipath-nested-Miller compensation. If $g_{m41} = g_{m3}$ and $g_{m42} = 2g_{m3}$, the resulting simplified numerator is

$$N(s) = 1 + s\frac{C_{M2} + C_{gs2}}{g_{m2}} + s^2\left[\frac{(C_{gs2})(C_{M1} + C_{be1}) + (C_{gs2} + C_{M2})(C_{be1})}{g_{m1}g_{m2}}\right] \qquad (4\text{-}91)$$

which contains the multipath zero and another zero in the left-half plane which is positioned at a relatively high frequency.

4.7 Dynamic distortion

To explain the high-frequency behavior, the simplified circuit diagram of a two-stage bipolar amplifier as depicted in Fig.4-29a is used. The amplifier is frequency compensated using a Miller capacitor C_{M1}. The discussion of high frequency distortion

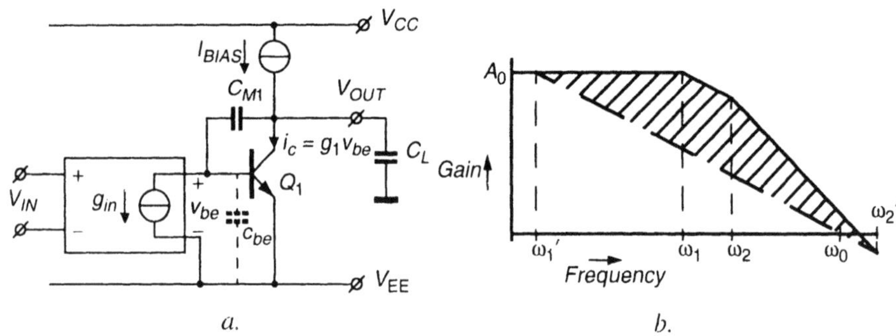

Fig. 4-29 *Simplified model of two-stage amplifier with Miller capacitor C_{M1} which mainly determines the high frequency behavior (a) and frequency response of output stage with and without frequency compensation (b). The dashed area represents the gain that is lost for linearizing the overall system but it is reused to linearize the output stage.*

can be divided into three parts. First of all, the Miller capacitor induces a different distortion mechanism than at dc. Secondly, the switch over from one output transistor to the other has to be taken into account. And, thirdly, the circuit that controls the quiescent current, the class-AB control circuit, should be considered.

At high frequencies, the gain is reduced by inserting the Miller capacitor as indicated by the dashed area in Fig.4-29b. Fortunately, the gain represented by this dashed area is not thrown away but used as loop gain in the feedback loop created by the Miller capacitor. Hence, the transfer function is linearized by C_{M1}. If we take a closer look at the circuit shown in Fig.4-29a, we see that the output voltage is not only determined by C_{M1} but also by the voltage across the base-emitter junction of the output transistor. Unfortunately, the high-frequency behavior of the output stage can only accurately be described by nonlinear differential equations which are difficult to solve. However, by using the assumption that the frequency of interest is well below the unity-gain frequency, the complexity of the problem can be much reduced. In that case, the signal current flowing through the capacitors C_{M1} and C_L is very small compared to the signal current through the load resistor R_L. Thus, the output voltage v_{out} can be expressed as a function of the input voltage v_{in} and the base-emitter voltage using the small-signal equation

$$v_{out} = \frac{g_{in}}{j\omega C_{M1}} v_{in} + v_{be} = \frac{g_{in}}{j\omega C_{M1}} v_{in} + \frac{v_{out}}{R_L} \cdot \frac{1}{g_1} \quad (4\text{-}92)$$

rewriting gives the open loop transfer function for high frequencies

$$\frac{v_{out}}{v_{in}} = \left(1 - \frac{1}{g_1 R_L}\right) \frac{g_{in}}{j\omega C_{M1}} \quad (4\text{-}93)$$

Thus, the transfer function for higher frequencies is not only determined by the Miller capacitor but also by the internal loop gain $g_1 R_L$ of the Miller feedback loop. In order to obtain low distortion at high frequencies, the Miller capacitor should be of high quality with very low voltage dependency, and the loop gain $g_1 R_L$ should be as large as possible requiring high quiescent currents. In circuits designed for low power applications the quiescent current is, for example, of the order of $100\mu A$, yielding a loop gain of about 10 when driving a 1-$k\Omega$ load. At the extremes of the output signal swing, the bias current and transconductance of the output stage are very high so that the error term in equation (4-93) changes from 0.9 to nearly unity as a function of the output voltage. Therefore, the gain of the amplifier at higher frequencies contains a signal dependent variation of 10%. The actual distortion also depends on the gain which is frequency dependent. At a frequency of a factor ten lower than the unity-gain frequency we find a distortion of about 1%, while around the unity-gain frequency the distortion increases to as much as 5%!

The second kind of high-frequency distortion relates to the switch over from one output transistor to the other one in push-pull stages. Therefore, we should consider the push-pull stage as depicted in Fig. 4-30 for high frequencies. For relatively small signal voltages, the distortion is determined by matching the transfer functions through the two output transistors. Since the transfer function for high frequencies is determined by the Miller capacitors C_{M1} and C_{M2}, good matching of the Miller capacitors of both output transistors is important. Any remaining differences are flattened around the quiescent state by applying a class-AB characteristic with a smooth switch over. For larger signals the dynamics of the push-pull output stage come into play. Each output transistor must be given sufficient time to switch to the required output current otherwise slewing effects will cause large distortion. Again, a class-AB characteristic with a gradual cross-over should be used since it turns on the required transistor well before it is needed. Further improvement is obtained by applying class-AB characteristics which maintain a residual

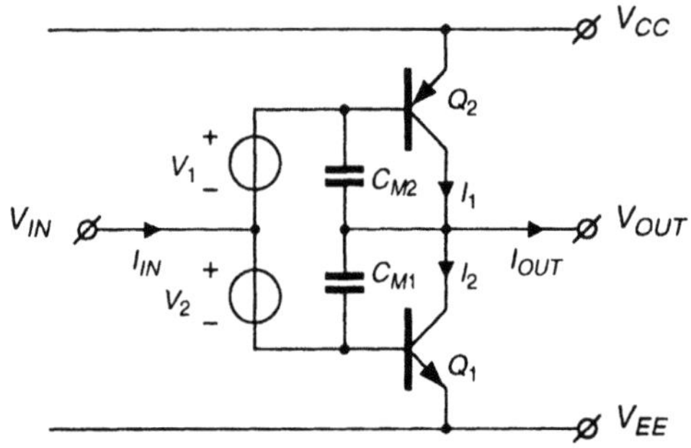

Fig. 4-30 *Simplified diagram of rail-to-rail output stage with frequency compensation and voltage sources which represent the bias control. The high frequency behavior is mainly determined by the two Miller capacitors.*

current in the inactive output transistor [24]. This measure reduces the signal swing at the base or gate of the output transistor to minimize the non-linear voltage across the Miller capacitors. This is especially important in bipolar stages where it takes a long time to charge the diffusion capacitor of a bipolar transistor that is completely switched off.

Finally, the dynamics of the class-AB control circuit can influence the behavior of the output stage. Therefore, much delay introduced by the class-AB control circuit cannot be allowed. To avoid delay, the control circuit should not interfere with the signal path, and the bandwidth should be larger than the bandwidth of the main amplifier.

The distortion at higher frequencies, or dynamic distortion is determined by the Miller capacitors and the loop-gain term $g_1 R_L$. Thus, both in bipolar and in CMOS technology, the transconductance of the output transistors dominates the distortion behavior. Again, matching the transfer functions of the output transistors and gradual switching over are important measures to minimize distortion.

To investigate the behavior of multi-stage nested-Miller circuits, we briefly consider the nested-Miller compensated three-stage amplifier shown in Fig. 4-21 and a mirrored-nested-Miller amplifier shown in Fig. 4-23 as examples. In the same way as with simple Miller compensation, we calculate the transfer function of the circuits to find the distortion behavior. The mid-frequency transfer function of the nested-Miller circuit is given by

$$\frac{v_{out}}{v_{in}} = \left(1 - \frac{1}{g_2 \beta_{F1} R_L}\right)\frac{g_3}{j\omega C_{M2}} \qquad (4\text{-}94)$$

The linearity of the Miller capacitor directly influences the high-frequency distortion. Further, the variation of β_{F1} as a function of the output current and the variation of g_{m2} as a function of the output current are important. But due to the higher gain, the distortion is much better than the mid-frequency distortion of a two-stage circuit. At higher frequencies, where the gain of the output stage reduces due to C_{M1}, the distortion of the three-stage circuit is of the same order as the distortion of a two-stage circuit.

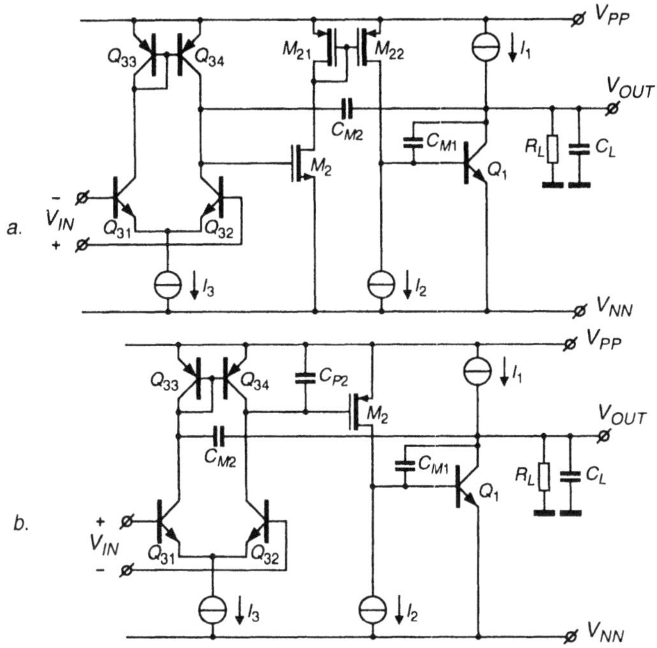

Fig. 4-31 *Example of a three-stage BiCMOS nested-Miller amplifier and a three-stage BiCMOS mirrored-nested-Miller amplifier to investigate mid-frequency distortion and programmability.*

The mid-frequency transfer function of the mirrored-nested-Miller amplifier is given by

$$\frac{v_{out}}{v_{in}} = \left(1 - \frac{1}{g_2 \beta_{F1} R_L \frac{g_{33}}{j\omega C_{P2}}}\right) \frac{g_3}{j\omega C_{M2}} \tag{4-95}$$

The gain inside the Miller loop is increased by the high-frequency gain of the current mirror, determined by the transconductance of Q_{33} and parallel capacitor C_{P2}. The result is an improved distortion.

4.8 Settling time

In Table 4-4 the time-domain parameters of several second-order responses are listed [3]. Clearly, the overshoot decreases with increasing damping ratio. The overshoot of the half-critical damping function is rather large. The Butterworth function achieves fast rise time t_r with acceptable overshoot of about 4%. Only if settling time is important, the Bessel response is advantageous. The settling time to 1% $t_{1\%}$ is slightly faster than the settling of the Butterworth function. But the settling to 0.1% $t_{0.1\%}$ is considerably faster for the same second-pole frequency and therefore the same power consumption! In all cases the single-pole function achieves excellent performance; but a single-pole response can only be obtained using a single-stage topology, or by shifting a second pole to very high frequencies which costs large amounts of power. Concluding, the best response type for a general purpose amplifier where unity-gain frequency and settling time are both important, is the Butterworth function. For amplifiers where accurate settling is the parameter of prime importance, the Bessel function should be implemented. It should be noted that the exact response function is only obtained using unity-gain feedback. For higher gain, the response slowly changes into a single-pole function.

Table 4-4 *Overview of time-domain parameters of second order functions*

Function type	ζ	t_r		$t_{1\%}$		$t_{0.1\%}$		**Over-shoot**
Half-critical damping	$\frac{1}{2}$	$1.64\tau_n$	$1.64\tau_2$	$8.78\tau_n$	$8.78\tau_2$	$12.7\tau_n$	$12.7\tau_2$	16.3%
Butterworth	$\frac{1}{2}\sqrt{2}$	$2.15\tau_n$	$3.04\tau_2$	$6.59\tau_n$	$9.32\tau_2$	$10.2\tau_n$	$14.5\tau_2$	4.32%
Bessel	$\frac{1}{2}\sqrt{3}$	$2.73\tau_n$	$4.74\tau_2$	$4.66\tau_n$	$8.08\tau_2$	$5.34\tau_n$	$9.25\tau_2$	0.43%
Coincident poles	1	$3.36\tau_n$	$6.72\tau_2$	$6.64\tau_n$	$13.3\tau_2$	$9.23\tau_n$	$18.5\tau_2$	0%
Single pole		$2.20\tau_n$	$2.20\tau_0$	$4.61\tau_n$	$4.61\tau_0$	$6.91\tau_n$	$6.91\tau_0$	0%

The next question is how the order of the system, or the number of stages, influences the performance. To investigate the influence of the number of stages, the performance of the Butterworth function up to the order of four is listed in Table 4-5 [3]. As we see, the overshoot and the settling time increase as a function of the order. If we also take into account the fact that when nested-Miller compensation is used the bandwidth reduces as a function of the number of stages, we again conclude that the number of stages should be as small as possible.

Table 4-5 *Parameters of Butterworth functions as a function of order n.*

n	Frequency-domain parameters		Time-domain parameters			
	ω_n	φ_m [°]	t_r	$t_{1\%}$	$t_{0.1\%}$	Overshoot
1	ω_0	90	$2.20\tau_n$	$6.64\tau_n$	$6.91\tau_n$	0%
2	$\sqrt{2}\,\omega_0$	63.4	$2.15\tau_n$	$6.59\tau_n$	$10.2\tau_n$	4.3%
3	$2\omega_0$	60.3	$2.29\tau_n$	$9.42\tau_n$	$13.5\tau_n$	8.1%
4	$2.61\omega_0$	60.7	$2.43\tau_n$	$12.0\tau_n$	$16.4\tau_n$	10.9%

The above derived results assume ideal second-order systems without the influence of pole-zero doublets and large-signal slewing effects. As already discussed in section 4.6.4, the existence of a pole-zero doublet results in a slow settling component [25] in the transient response with settling time [26]

$$T_S = \frac{\ln\left(\dfrac{S_d\omega_d}{a\omega_0}\right)}{\omega_d} \tag{4-96}$$

It is not possible to derive an equation that includes the ideal small-signal behavior and the slow-settling behavior. However, by comparing the settling time of the slow settling component with the settling of the ideal response, the influence of the slow-settling component can be estimated. As an example, consider the settling time of a three-stage multipath nested-Miller compensated amplifier, and let us compare it to the settling time of a three-stage nested-Miller circuit. Using Table 4-4, an ideal multipath nested Miller amplifier with exact matching of the pole and the zero would settle to 0.1% within $7.21\tau_0$. If the frequency of the pole-zero doublet of the amplifier is equal to $^1/_2\omega_0$, the slow-settling component settles to 0.1% within $2.3\tau_0$, if the pole-zero spacing S_d is 2%. Thus, the slow-settling component settles much faster than the ideal amplifier so that the settling time is only slightly increased by the pole-zero doublet. If the pole-zero spacing is 10%, the slow settling component settles within $7.8\tau_0$ which is of the same order as the ideal response. Therefore, the settling time is considerably increased. The nested-Miller amplifier settles to 0.1% within $6.75\tau_0$. However, since its unity-gain frequency is a factor two smaller than the multipath amplifier, the multipath amplifier settles faster even with a 10% pole-zero mismatch. Concluding, the multipath compensation technique is also beneficial regarding settling time because of the reduction of the order of the system in spite of the introduction of a pole-zero doublet. This is at least true if the frequency of the doublet is not too small, of the order of $^1/_2\omega_0$.

For large signals the settling speed is reduced by the finite slew rate of amplifiers. The slew rate is caused by the application of differential input stages, as shown in the basic topology of a bipolar amplifier in Fig. 4-32a. The differential stage can only deliver a limited amount of current equal to the tail current source I_N to the compensation capacitor. Therefore, the output voltage can only rise at a limited rate which is called slew rate. The slew rate of a Miller compensated amplifier is therefore given by [11]

$$S = \frac{I_N}{C_{M,(N-1)}} = \frac{I_N}{g_{m,N}}\omega_0 \qquad (4\text{-}97)$$

where I_N is the bias current of the input stage and $C_{M,(N-1)}$ the outer Miller capacitor, as shown in Fig. 4-32a. In the case of bipolar technology, the relation between slew rate and unity-gain frequency is fixed

$$S = 2\frac{kT}{q}\omega_0 = 4\pi\frac{kT}{q}f_0 \approx \frac{1}{3.2}f_0 \qquad (4\text{-}98)$$

In order to calculate the delay introduced by the slewing effect, we consider the influence of the slewing on the step response of a first-order system. The ideal first-order step response is shown in Fig. 4-32b, together with the slew-rate limited response. Initially, the rise of the voltage of the unlimited response is high. In that case, the limited response rises at a constant rate determined by S. After some time t_1, the rise of the voltage of the unlimited response equals the slew rate. At that specific voltage $V(t_1)$, the limited response enters the small-signal regime and finally commences with its small-signal settling at time t_2. The delay due to slewing is now given by

$$\Delta t_S = t_2 - t_1 = \frac{1}{2\pi f_0}\ln\left(\frac{g_m V_S}{2I} - 1 - \ln\frac{g_m V_S}{2I}\right) \qquad (4\text{-}99)$$

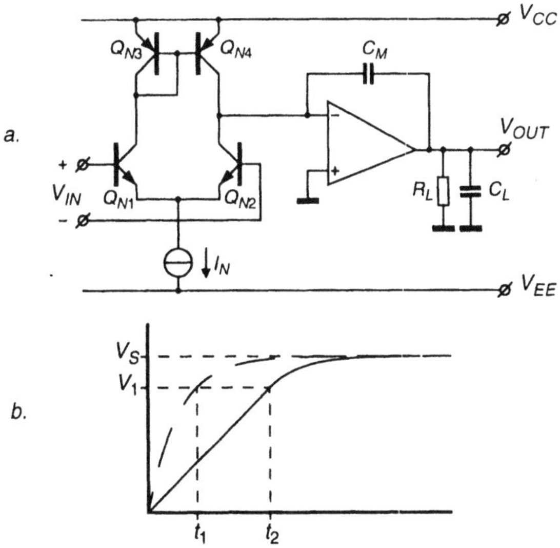

Fig. 4-32 *Basic bipolar amplifier configuration for determining slew-rate limitation (a) and effect on the step response of a single pole amplifier (b). Dashed line indicates ideal response and solid line indicates response limited by slew rate.*

which reduces in bipolar technology to

$$\Delta t_{S,\,Bip} = \frac{1}{2\pi f_0}\ln\left(\frac{V_S}{2V_T} - 1 - \ln\frac{V_S}{2V_T}\right) \qquad (4\text{-}100)$$

The effect on higher-order systems can be approximated by the above results.

4.9 Design for programmability

A final important property of an amplifier design is the possibility of scaling so that certain parameters can be influenced by the user, usually by changing the bias current. The most important parameter that is easily influenced by the bias current is the bandwidth. An amplifier of which the bandwidth can be changed by modifying the bias current is called programmable [27]. The bias current can be changed either by modifying an external bias current or by changing the PTAT resistor of the applied PTAT current source. This presents hardly any problems in pure CMOS designs where all stages are biased in the same region. Also, in pure bipolar designs, scaling is easily performed. However, resistors used to bias cascodes need to be scaled by the same factor in order to bias the current sources below the cascodes properly. It is therefore clever to place these resistors next to the PTAT resistor so that they can be easily stretched together. More problems arise in BiCMOS technology or in CMOS technology if some stages are biased in strong inversion while other stages are biased in weak inversion. Let us, for example, consider the three-stage amplifier shown in Fig. 4-31a. The pole at the output is determined by the transconductance of Q_1 and the load capacitor C_L, and scales proportional to the bias current. The unity-gain frequency is determined by the transconductance of input stage Q_{31}, Q_{32} and Miller capacitor C_{M2}, and also scales proportional to the bias current. If the intermediate stage M_2 is biased in strong inversion, the pole of the intermediate stage scales proportional to the square root of the bias current. Therefore, the relative placement of the poles changes if we only change the bias current. For correct scaling, the transconductance of M_2 has to be further scaled by a factor equal to the square root of the bias-current scaling. This can be done, for example, by changing the W/L ratio of M_2 by the same factor as the bias current. In the case of a cascoded-nested-Miller circuit or a mirrored-nested-Miller circuit, as depicted in Fig. 4-31b, an additional possibility to correctly scale the circuit is to change C_{P2} by the square root of the scaling factor. Using these measures, the performance of BiCMOS amplifiers can be correctly scaled.

4.10 Conclusions

To obtain sufficient gain, at least two stages, but in most cases, three stages are required. The task of the frequency compensation is to shape the frequency response of the two-stage and three-stage amplifiers according to the Butterworth response. We have seen that Miller compensation is preferred because it is insensitive to parameter variations. For three-stage designs, nested-Miller compensation can be applied. However, the price to be paid for the insensitivity is a reduction in bandwidth. The bandwidth reduction can be prevented in two-stage amplifiers using cascoded Miller compensation. However, the cascoded technique is sensitive to changes in the output-stage bias current so that it cannot be used in combination with class-AB output stages. A compromise can be obtained using nested-cascoded-Miller compensation but this technique requires quite some design effort.

A modification of nested-Miller compensation is mirrored-nested-Miller compensation which allows simplification of the nested structure because it eliminates a current mirror. Finally, the multipath technique was introduced which can be used to improve the bandwidth reduction of three-stage nested-Miller schemes. The three-stage version of multipath-nested-Miller compensation achieves the same bandwidth as two-stage Miller compensation while obtaining more gain. A drawback of multipath techniques, however, is that a pole-zero doublet occurs which is, fortunately, well controlled.

4.11 References

[1] K. Bult, G.J.G.M. Geelen, "A fast settling CMOS op amp for SC-circuits with 90-dB DC gain", *IEEE J. Solid-State Circuits*, vol. SC-25, no. 12, pp. 1379-1383, Dec. 1990.

[2] R. Hogervorst, S.M. Safai, J.P. Tero, J.H. Huijsing, "A Programmable Power-Efficient 3-V CMOS Rail-to-Rail Opamp with Gain Boosting for Driving Heavy Resistive Loads", *Proc. IEEE International Symposium on Circuits and Systems*, Seattle, USA, pp. 1544-1547, April 30 - May 3, 1995.

[3] E.M. Cherry, D.E. Hooper, *Amplifier Devices and Low-pass Amplifier Design*. New York: Wiley, 1968.

[4] H.W. Bode, *Network Analysis and Feedback Amplifier Design*. New York: Van Nostrand, 1949.

[5] H. Nyquist, "Regeneration Theory", *Bell Systems Technical Journal*, Jan. 1932.

[6] E.M. Cherry, D.E. Hooper, *Amplifier Devices and Low-pass Amplifier Design*. New York: Wiley, 1968.

[7] R.G.H. Eschauzier and J.H. Huijsing, *Frequency compensation techniques for low-power operational amplifiers*. Dordrecht, The Netherlands: Kluwer Academic Publishers, 1995.

[8] M.J. Fonderie, J.H. Huijsing, *Design of Low-voltage Bipolar Operational Amplifiers*. Boston: Kluwer, 1993.

[9] K.J. de Langen, R.G.H. Eschauzier, and J.H. Huijsing, "A 1-GHz Class-AB Operational Amplifier with Multipath Nested Miller Compensation for 76-*dB* Gain", *IEEE J. Solid-State Circuits*, vol. SC-32, no. 4, pp. 488-498, April 1997.

[10] K.J. de Langen, J.H. Huijsing, "Compact Low-Voltage Power-Efficient Operational Amplifier Cells for VLSI", *IEEE J. Solid-State Circuits*, vol. SC-33, no. 10, pp. 1482-1496, Oct. 1997

[11] J.E. Solomon, "The Monolithic Op Amp: A Tutorial Study", *IEEE J. Solid-State Circuits*, Vol. SC-3, No. 6, pp. 314-332, Dec. 1974.

[12] R.J. Reay, G.T.A. Kovacs, "An Unconditionally Stable Two-Stage CMOS Amplifier, *IEEE J. Solid-State Circuits*, vol. SC-30, no. 5, pp. 591-594, May 1995.

[13] R.G.H. Eschauzier, J.H. Huijsing, "An Operational Amplifier with Multipath Miller Zero Cancellation for RHP Zero Removal", in *Proc. ESSCIRC* 1993, Editions Frontières, Gif-sur-Yvettes, France.

[14] J.H. Huijsing, R.G.H. Eschauzier, "Amplifier arrangement with multipath miller zero cancellation", U.S Patent No. 5,485,121, Jan. 16, 1996.

[15] B.K. Ahuja, "An Improved Frequency Compensation Technique for CMOS Operational Amplifiers, *IEEE J. Solid-State Circuits*, vol. SC-18, no. 6, pp. 629-633, Dec. 1983.

[16] R. Hogervorst, J.H. Huijsing, *Design of Low-Voltage Low-Power Operational Amplifier Cells*. Dordrecht, The Netherlands: Kluwer, 1996.

[17] J.H. Huijsing, "Multistage amplifier with capacitive nesting for frequency compensation", U.S. Patent No. 4,559,502, Dec. 17, 1985.

[18] J.H. Huijsing, D. Linebarger, "Low-Voltage Operational Amplifier with Rail-to-Rail Input and Output Ranges", *IEEE J. Solid-State Circuits*, Vol SC-20, No. 6, pp. 1144-1150, Dec. 1985.

[19] J.H. Huijsing and M.J. Fonderie, "Multi-stage amplifier with capacitive nesting and multi-path forward feeding for frequency compensation", U.S Patent No. 5,155,447, Oct. 4, 1992.

[20] M.J. Fonderie and J.H. Huijsing, "Operational Amplifier with 1-V Rail-to-Rail Multipath-Driven Output Stage, *IEEE J. Solid-State Circuits*, vol. SC-26, no. 12, pp. 1817-1824, Dec. 1991.

[21] R.G.H. Eschauzier, L.P.T. Kerklaan, and J.H. Huijsing, "A 100-MHz 100-dB Operational Amplifier with Multipath Nested Miller Compensation Structure", *IEEE J. Solid-State Circuits*, vol. SC-27, no. 12, pp. 1709-1717, Dec. 1992.

[22] B.Y. Kamath, R.G. Meyer, P.R. Gray, "Relationship between Frequency Response and Settling Time of Operational Amplifier", *IEEE J. Solid-State Circuits*, vol. SC-9, pp. 347-352, no. 6, Dec. 1974.

[23] F. You, S.H.K. Embabi, E. Sánchez-Sinencio, "Multistage Amplifier Topologies with Nested G_m-C Compensation", *IEEE J. Solid-State Circuits*, vol. SC-32, no. 12, pp. 2000-2011, Dec. 1997.

[24] E. Seevinck, W. de Jager, P. Buitendijk, "A Low-Distortion Output Stage with improved stability for monolithic power amplifiers", *IEEE J. Solid-State Circuits*, Vol. SC-23, pp. 794-801, June 1988.

[25] B.Y. Kamath, R.G. Meyer and P.R. Gray, "Relationship Between Frequency Response and Settling Time of Operational Amplifiers", *IEEE J. Solid-State Circuits*, vol. SC-9, no. 6, pp. 347-352, Dec. 1974.

[26] R.G.H. Eschauzier and J.H. Huijsing, *Frequency compensation techniques for low-power operational amplifiers*. Dordrecht, The Netherlands: Kluwer Academic Publishers, 1995.

[27] R.G.H. Eschauzier, R. Hogervorst, J.H. Huijsing, "A Programmable 1.5 V CMOS Class-AB Operational Amplifier with Hybrid Nested Miller Compensation for 120 dB Gain and 6 MHz UGF", *IEEE J. Solid-State Circuits*, vol. SC-29, no. 12, pp. 1497-1504, Dec. 1994.

Realizations

5

5.1 Introduction

Now that we have discussed all the main overall parameters, it is time to consider actual amplifier implementations in more detail. The amplifier topologies are described in subsequent sections classified according to the applied technology. In section 5.2 bipolar amplifier topologies are discussed: we will see that bipolar amplifiers due to the availability of the high performance NPN transistor, are very suited for wideband circuits and for amplifiers with high output current capability. Amplifiers that are suited for VLSI are presented in section 5.3. These CMOS amplifiers are very compact, making the best of a simple two-stage topology while being able to operate at low supply voltages. At extremely low supply voltages, however, three stages are required yielding a compact ultra-low-voltage design. The final group of amplifiers employs the possibilities of BiCMOS technology yielding powerful combinations of the previous discussed topologies. BiCMOS topologies are presented in section 5.4. Finally, conlusions are drawn in section 5.5.

5.2 Bipolar amplifier topologies

Due to the intrinsic high transconductance, the bipolar transistor is more efficient regarding bandwidth as a function of supply current. Therefore, bipolar amplifiers are especially useful for obtaining the highest bandwidth. Also, the required supply voltage for driving large output currents is small, thanks to the exponential relation between base-emitter voltage and collector current. This makes bipolar technology interesting for designing efficient amplifiers that can deliver large output current with a small standby current operating at low supply voltages. Wideband amplifiers are discussed in section 5.2.1. An efficient amplifier topology for driving high output currents is presented in section 5.2.2.

5.2.1 1-GHz operational amplifier

Multipath compensation techniques enable multi-stage circuits with virtually no bandwidth reduction. Therefore, multipath-nested-Miller compensation can be used in wideband amplifier design in order to obtain better accuracy than conventional local feedback techniques. But first we consider the basics of wideband amplifiers and design a two-stage amplifier. Thereafter a three-stage circuit is discussed that obtains sufficient gain with the low ohmic loads of the order of 50 Ohms which are used in wideband systems.

To obtain the highest possible bandwidth, only the device with the highest transit frequency, the NPN transistor, should be applied in the high-frequency signal path. Even in modern complementary processes, the NPN is a factor three faster than its PNP counterpart. Because of the fundamental difference in mobility of electrons and holes, this will not change in the future. Therefore, very wideband amplifiers should exploit an all-NPN topology [1]. For low frequencies, however, PNP transistors can be used which are required to implement the dc levelshifts between NPN stages.

A simplified schematic of a two-stage all-NPN opamp is shown in Fig. 5-1. It consists of an NPN input stage Q_3, Q_4, and an NPN output stage Q_1, Q_2. Apart from the fact that only NPN transistors are used, the high-frequency behavior benefits from the very simple topology. Because the output stage consists of a follower Q_1 and an inverter Q_2, the output stage can be directly driven by the differential input stage Q_3, Q_4, without a mirror, which would have added delay and parasitics. A practical implementation, however, gives rise to three main problems. First, the common-mode input voltage should be independent of the biasing voltages of the output stage and the output voltage. To achieve this, level-shifts V_{LS1},V_{LS2} are necessary between input stage and output stage. Of course, the level-shift circuits should not introduce much delay and parasitic capacitance. Secondly, the output stage is asymmetric, one output transistor being

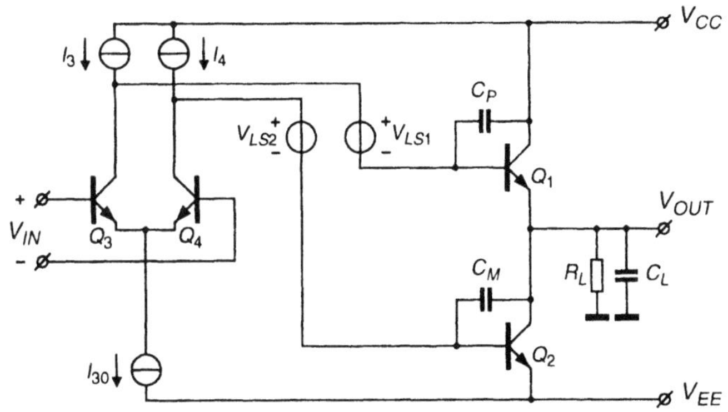

Fig. 5-1 *Basic two-stage all-NPN topology.*

connected as a follower and the other output transistor as an inverter. As demonstrated in chapter four, the behavior can easily be balanced. But does this also function properly at very high frequencies?

The implementation of the level-shift circuit is depicted in Fig. 5-2. In order to realize a transfer function form input current I_{IN} to output current I_{OUT} that is flat over a wide frequency range, a PNP cascode Q_1 is combined with an NPN cascode Q_2. The PNP cascode is applied for low-frequency signals while the NPN cascode is used at high frequencies as determined by RC all-pass network R_{F1}, C_{F1}. The all-pass network splits the input signals into a low-frequency component I_{LF} and a high-frequency component I_{HF}. The low-frequency component flows via R_{F1} and Q_1 to the output node, while I_{HF} flows via C_{F1} and Q_2 to the output. If the turnover frequency $\omega_{TO} = 1/R_{F1}C_{F1}$ is much

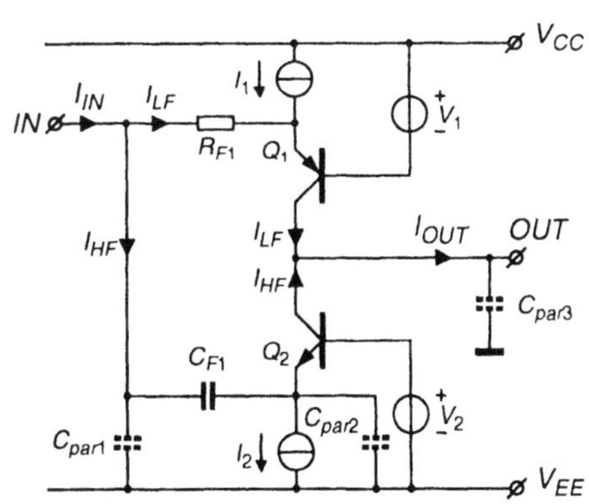

Fig. 5-2 *Wideband level-shift implementation.*

lower than the transit frequency of the PNP transistor, the response of the circuit is flat up to the transit frequency of the NPN transistor [1] without pole-zero doublets. Unfortunately, several non-idealities influence the behavior of the level shift. For low frequencies, the transfer function is attenuated by the current gain of the PNP transistor which is considerably smaller than the current gain of the NPN transistor. For high frequencies, the transfer function is influenced by parasitic capacitances C_{par1}- C_{par3}, which are shown in Fig. 5-2. C_{Par1} consists of the capacitive loading of the input stage and attenuates the high-frequency transfer function. The parasitic junction capacitance of C_{F1} contributes to C_{par1} or C_{par2}, depending on the orientation of C_{F1}. Capacitance C_{par2} is connected across the base-emitter capacitance of Q_2 and therefore reduces the bandwidth of the NPN cascode. Capacitance C_{Par3} comprises the substrate capacitances of the level-shift transistors and loads the output stage.

The output transistors with their parasitics are shown in Fig. 5-3. The parasitic capacitances C_{par31} and C_{par32} are the capacitive loading of the level-shift circuits that drive Q_1 and Q_2, respectively. As discussed in chapter four, the output stage is balanced

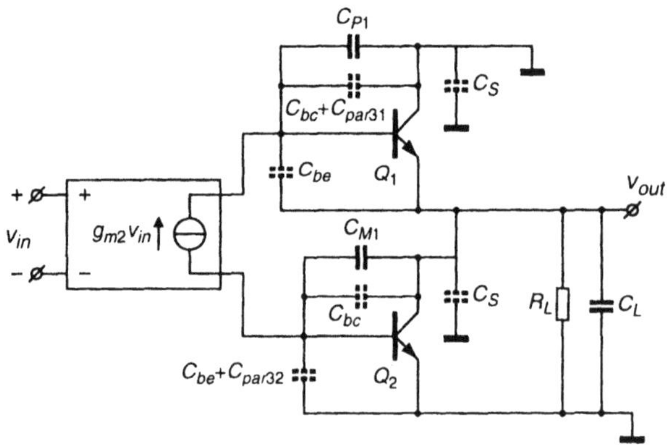

Fig. 5-3 *Combination of output transistors.*

for low-frequencies if the current gain β_F of the output transistors is equal, and if they are driven by ideal current sources. Further, the mid-frequency transfer function is balanced if the capacitances connected across the base-collector junctions of both output transistors are equal. In order to investigate the high-frequency behavior, the transfer functions of the follower and the inverter are calculated. The transfer function of the follower is given by

$$\frac{V_{out}}{V_{in}} = \frac{g_2\beta_F R_L\left(1+s\frac{C_{be}}{g_1}\right)}{1+sr_{be}[(C_{be}+C_{P1})+\beta_F R_L C_{P1}]+s^2 r_{be}R_L(C_{be}C_L+C_{be}C_{P1}+C_L C_{P1})} \tag{5-1}$$

where g_1 is the transconductance of the output transistors, g_2 the transconductance of the input stage, β_F the current gain of the output transistors, R_L the load resistance, r_{be} the small-signal base-emitter resistance of the output transistors, C_{be} the base-emitter capacitance of the output transistors, and C_L the load capacitance. C_{P1} also includes the base-collector capacitance C_{bc} and C_{par31}. The transfer function of the inverter is found as

$$\frac{V_{out}}{V_{in}} = \frac{-g_2\beta_F R_L\left(1-s\frac{C_{M1}}{g_1}\right)}{1+s[r_{be}(C_{be}+C_{M1})+\beta_F R_L C_{M1}]+s^2 r_{be}R_L(C_{be}C_L+C_{be}C_{M1}+C_L C_{M1})} \tag{5-2}$$

where C_{M1} is the Miller capacitor which includes C_{bc}. Capacitance C_{be} also includes C_{Par32}.

Concluding, when both transistors are biased at the same current and all capacitors of the follower have been made equal to the capacitors of the inverter, the only difference that remains is that the zero of the follower is positioned in the left half-plane, while the zero of the inverter is positioned in the right half-plane. Due to the phase shift of the right half-plane zero, the bandwidth of the inverter is smaller than the bandwidth of the follower.

The operating conditions of wideband amplifiers require special care. The load is mainly determined by the low resistive impedance while the load capacitor is relatively small. On the other hand, the base-emitter capacitor is relatively large because of the high bias currents. Thus, all capacitors are of the same order. These special circumstances require different approximations than the approximations derived in chapter four. Further, because we now consider a balanced output stage consisting of two transistors several factors two appear in the equations. Therefore, we briefly review the main equations that determine the frequency compensation. If C_{M1} and C_{P1} are equal, the unity-gain frequency ω_0 for both output transistors is found as

$$\omega_0 = \frac{g_2}{C_{M1}} \cdot \frac{1}{1 + \frac{1}{g_1 R_L}\left(1 + \frac{C_{be}}{C_{M1}}\right)} \qquad (5\text{-}3)$$

As we see, the unity-gain frequency is influenced by the loop-gain term $g_1 R_L$. Because of the low resistive load of 50 ohms, this term can be quite small. Therefore, a large output-stage bias current is needed of at least several milliamps to assure that the unity-gain frequency is accurately determined by C_{M1} and the input-stage transconductance. Then the correction term is close to one, and the bandwidth is accurately set by g_2 and C_{M1}

$$\omega_0 = \frac{g_2}{C_{M1}} \qquad (5\text{-}4)$$

When the output stage operates near the quiescent state, both output transistors work together. The transfer function of the combination has three poles. However, when the follower and inverter transistors are balanced, mainly by adjusting C_{P1} so that $C_{P1}+C_P$ equals C_{M1}, one pole is exactly cancelled by a zero at

$$\omega_{PZ} = \frac{1}{r_{be}(C_{M1} + C_{be})} \qquad (5\text{-}5)$$

And the transfer function is found as

$$\frac{V_{out}}{V_{in}} = \frac{2g_2\beta_F R_L\left(1 + s\frac{C_{be} - C_{M1}}{g_1}\right)}{1 + sr_{be}[(C_{be} + C_{M1}) + 2\beta_F R_L C_{M1}] + s^2 r_{be} R_L (C_{be}C_L + 2C_{be}C_{M1} + C_L C_{M1})} \qquad (5\text{-}6)$$

The zero disappears when C_{be} equals C_{M1}. Then, the current through C_{be} compensates the feedforward current through C_{M1}, thus removing the right half-plane zero much in the same way as with multipath Miller zero cancellation [2]. Because of the high bias current in the output stage, C_{be} is of the order of C_{M1} and the zero is shifted to very high frequencies. In cases where C_{be} is too small, a resistor can be inserted in series with C_{M1} to shift the zero to high frequencies. This has the disadvantage that the position of the zero changes as a function of the output current. However, due to the large quiescent current in the output stage, the inserted resistor is relatively small. Therefore, even at high output currents, the zero is situated at very high frequencies.

If $g_1 R_L$ is large, the unity-gain frequency of the combination is equal to that of the separate output transistors, as given by equation (5-4). Thus, the signal behavior in quiescent state is the same as in the situation where high output currents are handled assuring good high-frequency linearity.

The Miller capacitor C_{M1} which is used to split the two poles of the two-stage opamp apart [3,4], has to be chosen such that the unity-gain frequency is below the second pole. The second pole of the combination is given by

$$\omega_{p2} = \frac{g_1}{C_{be} + \frac{1}{2}C_L} \tag{5-7}$$

In order to obtain a Butterworth response, the value of the Miller capacitor should be

$$C_{M1} = 2\frac{g_2}{g_1}\left(C_{be} + \frac{1}{2}C_L\right) \tag{5-8}$$

Because of the small load capacitor, we expect to find the maximum bandwidth at a considerable input-stage bias current. Therefore, it is worthwhile to consider the bandwidth as a function of the bias-current division. The bandwidth of the balanced output stage is given by

$$\omega_{B,M} = \sqrt{\frac{g_1 g_2}{\frac{1}{2}C_{be1}C_L + \frac{1}{2}C_L C_{M1} + C_{be1}C_{M1}}} \tag{5-9}$$

In the same manner as in chapter five, a current division parameter α is defined and the current dependency of the base-emitter capacitance is introduced. The bandwidth as a function of α is plotted in Fig. 5-4. As an example, a total current I_t of 7 *mA* and a load capacitance of 1 *pF* is used. The maximum is around α=0.35, which means that 35% of the total supply current should be used to bias the input stage. We need about 5 *mA* bias current in the output stage to obtain sufficient loop gain in the Miller loop, and about 1 *mA* bias current per transistor in the input stage, because of noise considerations. Fortunately, this is close to the optimum. The resulting unity-gain frequency is 3.4 *GHz* using a Miller capacitor of 0.9 *pF*.

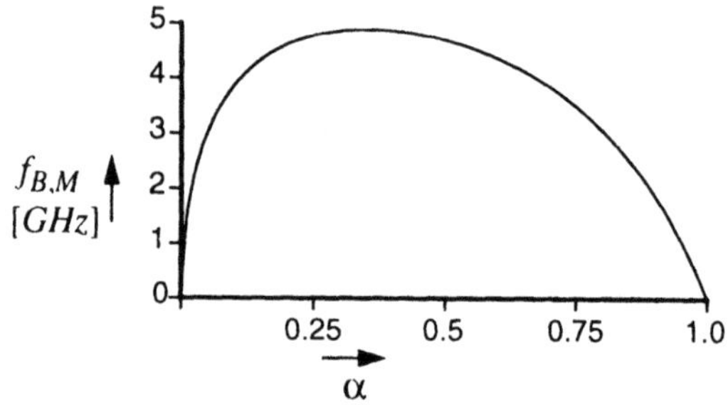

Fig. 5-4 *Bandwidth $f_{B,\,M}$ of a Miller compensated opamp as a function of the fraction α of the total current I_t used as bias current I_2 of the input stage.*

A final subject in the design of the two-stage circuit is the biasing of the output transistors. When we want to be able to handle signals with an amplitude of a few volts, considerable current should be delivered to the 50 Ω load. If the amplitude is, for instance, 2 *V*, a maximum current of 40 *mA* is required! Since we only need 5 *mA* bias current in the output stage a class-AB biasing would not be efficient. Therefore, a class-AB bias control is implemented using a feedback loop with minimum current control [5], as demonstrated in chapter three, section 3.6. The complete two-stage wideband amplifier including the class-AB control is shown in Fig. 5-5. The bias current of output

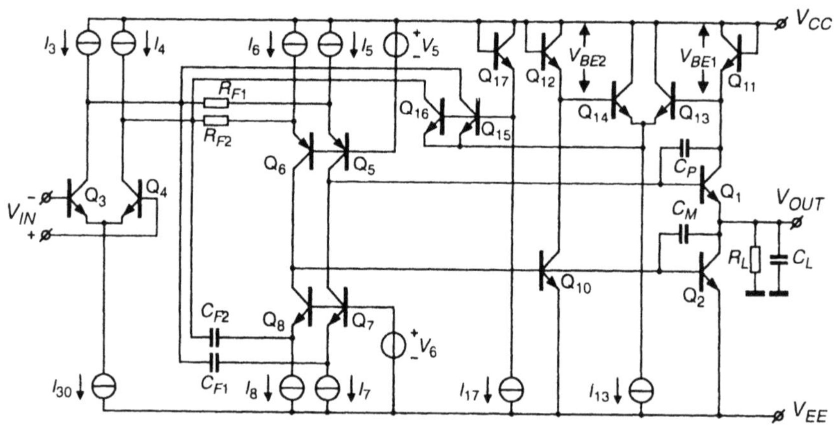

Fig. 5-5 *Two-stage 1-GHz opamp.*

transistor Q_1 is measured by placing diode-connected transistor Q_{11} in series with the collector lead of Q_1. Thus, a voltage V_{BE1} is created across Q_{11} that is proportional to the base-emitter voltage of Q_1. The bias current of output transistor Q_2 is measured by placing the base-emitter junction of Q_{10} in parallel with the base-emitter junction of Q_2. The collector current of Q_{10} generates a voltage V_{BE2} across diode-connected transistor Q_{12}. The emitter voltage of Q_{11} and Q_{12} drive decision-pair transistors Q_{13} and Q_{14}. The decision pair decides which of the two voltages V_{BE1}, V_{BE2} is the smaller because the output, the common emitter of the decision pair, is driven by the higher of the two input voltages which corresponds to the smaller of the two voltage drops V_{BE1}, V_{BE2}. The smaller of the two voltages V_{BE1}, V_{BE2} corresponds to the smaller of the output-stage bias currents and, therefore, the biasing of the smaller output-stage current. The control loop is completed by transistors Q_{15} and Q_{16} which are driven by the decision pair, and generate two equal currents that are subtracted from the inputs of the level-shift circuits R_{F1}, Q_5, C_{F1}, Q_7 and R_{F2}, Q_6, C_{F2}, Q_8. Thus, the control loop adjusts the input current of the output stage via the level-shift circuits in such a way that the lower of the collector currents of both output transistors stays above a certain minimum, avoiding cutoff of the output transistors. By connecting the collectors of Q_{15} and Q_{16} to the inputs of the level shifts, the class-AB control takes advantage of the high bandwidth of the level-shift circuits guaranteeing a stable feedback loop. The all-NPN circuit with minimum-current control assures excellent linearity from low frequencies up to very high frequencies.

Unfortunately, the dc-gain of the two-stage opamp from is only

$$A_0 = 2g_2\beta_F R_L \tag{5-10}$$

which amounts to about 44 *dB* using 1 *mA* bias current for the input transistors, a load resistance of 50 Ω, and a current gain β_F of 80. This is not acceptable for a general-purpose wideband operational amplifier. Therefore, a third stage is needed. By adding this stage between input stage and the level-shift stage, the noise contribution of the class-AB control transistors is reduced by the gain of this intermediate stage, improving the noise performance by a factor of $\sqrt{2}$.

Addition of the third stage increases the gain to about 80 *dB* while a multipath structure is applied to maintain high bandwidth [6,7,8]. The three-stage wideband operational amplifier with multipath-nested-Miller compensation is depicted in Fig. 5-6 [9]. The standard input stage Q_{23}, Q_{24} drives intermediate stage Q_3, Q_4. The intermediates stage Q_3, Q_4, the output stage Q_1, Q_2, the class-AB control circuit Q_{10}-Q_{17}, the level shifts Q_5-Q_7, R_{F1}, R_{F2}, C_{F1}, C_{F2}, and the output-stage compensation capacitors C_{M1}, C_{P1} are all connected in the same way as in the two-stage circuit. Capacitor C_{M2} implements the nested-Miller compensation of the three-stage topology. Capacitor C_{P2} is added to provide the high-frequency grounding reference for the non-inverting input of the internal two-stage amplifier Q_3, Q_4, Q_1, Q_2. A second input-stage Q_{25}, Q_{26} implements

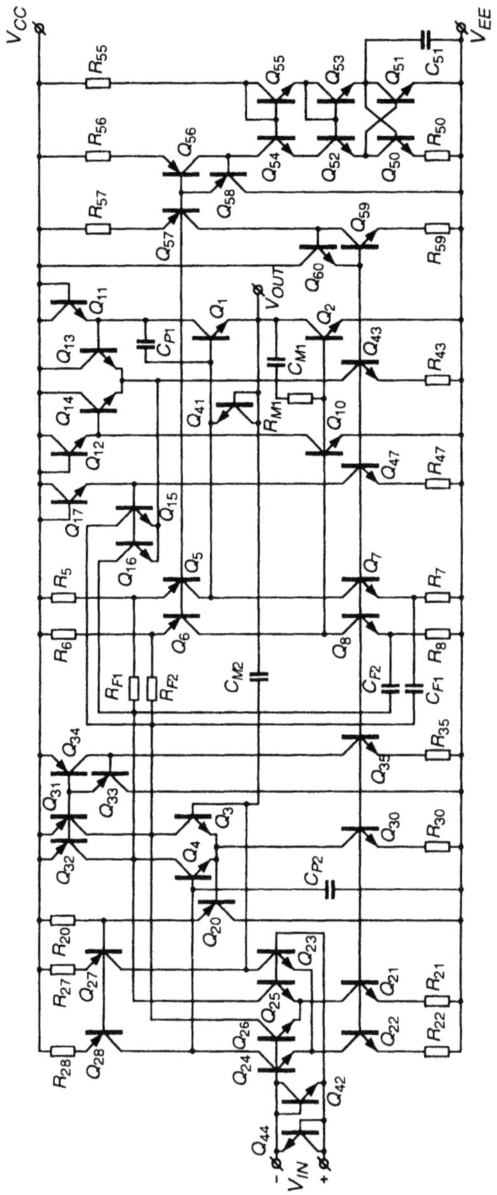

Fig. 5-6 *Complete three-stage 1-GHz operational amplifier with multipath-nested-Miller compensation.*

an independent feedforward path which bypasses intermediate stage Q_3, Q_4 and directly drives the output transistors via the level shifts. By using this structure, high gain of the three-stage topology is combined with high bandwidth of the two-stage circuit, as was proven in chapter four. The value of the multipath zero in the complete push-pull circuit is

$$\omega_z = \frac{1}{r_{be2}\left(C_{be2} + \frac{1}{2}C_{M2}\right)} \tag{5-11}$$

When this zero is positioned correctly, the feedforward path takes over exactly at the second pole and extends the -20 dB/dec. slope towards the third pole. This is obtained when the unity-gain frequency of the low-frequency and high-frequency transfer function are equal [8] yièlding the approximated requirement

$$\omega_0 = 2\frac{g_{31}}{C_{M2}} = \frac{g_{32}}{C_{M1}} \tag{5-12}$$

The mid-frequency gain of the feedforward path A_{FF} which determines the bandwidth reduction and the frequency of the pole-zero doublet as discussed in section 4.6.4 is

$$A_{FF} = \frac{g_{32}}{g_2}\left(1 + 2\frac{C_{be2}}{C_{M2}}\right) \approx \frac{g_{32}}{g_2} \tag{5-13}$$

The bandwidth $\omega_{B,\,MNM}$ of the multipath-nested-Miller compensated circuit is given by

$$\omega_{B,\,MNM} \cong \sqrt{\frac{g_1 g_{32}}{C_{be1}C_L + \frac{1}{2}C_L C_{M1} + C_{be1}C_{M1}}} \tag{5-14}$$

which is almost identical to the bandwidth expression of the two-stage circuit. Therefore, the bandwidth remains high and the bias-current considerations are similar to the bias-current considerations of the Miller compensated circuit. Thus, the input stage in our example is still biased at 1 *mA* per transistor, and the output stage at 5 *mA*. At high frequencies, where the feedforward path takes over, the noise of the amplifier is determined by the feedforward input stage and the intermediate stage. Therefore, both the feedforward input stage and the intermediate stage need a high bias current. Although the bandwidth does not depend on the intermediate-stage transconductance g_2

in first order, the bandwidth reduces if A_{FF} is small [8], as discussed in section 4.6.4. On the other hand the settling behavior improves for low values of A_{FF} and the noise at higher frequencies improves because the take-over frequency of the feedforward path is increased. A compromise is found by biasing both the feedforward input stage and the intermediate stage at 1 mA per transistor. Altogether, the unity-gain frequency of the example is 2.8 GHz. The value of C_{M1} and C_{P1} is 1.1 pF and the value of C_{M2} and C_{P2} is 2.2 pF.

Since the source impedance in high-frequency systems is low, usually 50 Ω, the noise performance is dominated by voltage noise. The equivalent input noise voltage resistance of the opamp is dominated by that of the input pair

$$R_{eqv} = 2R_{B,\,23,\,24} + \frac{1}{g_{m,\,23,\,24}} + 2\frac{1}{g_{m,\,23,\,24}^{2} R_P} \tag{5-15}$$

where $R_{B,23,24}$ is the base resistance of the input stage Q_{23}, Q_{24}, $g_{m,23,24}$ is the transconductance of the input stage Q_{23}, Q_{24}, and R_p is the equivalent noise resistance of current sources I_{23} and I_{24}. The first two terms originate from the input stage, and the last term from the current sources I_{23} and I_{24}. The last term can be made small by applying emitter degeneration. In this design, we allow a noise voltage of 1.2 $nV/\sqrt{Hz}$. Since 1.2 $nV/\sqrt{Hz}$ corresponds to a resistance of only 90 Ω, a very low base resistance R_B and a very low emitter impedance $^1/g_m$ are necessary. Therefore, very large input transistors should be used to realize a base resistance of the order of 30 Ω, and a high collector bias current of the input transistors of the order of 1 mA is required to achieve an emitter resistance $^1/g_m$ of about 25 Ω at room temperature.

The feedforward input stage contributes to the noise at high frequencies. Although this seems to be less important, the affected frequency band can be very large. Therefore, the noise considerations also apply to the feedforward input stage and the take-over of the feedforward path should be at the highest frequency possible. Since the take-over is determined by A_{FF} as given by equation (5-13), the transconductance g_2 of the intermediate stage must be large. The input-referred noise characteristic of the multipath-nested-Miller compensated wideband amplifier illustrating the higher noise above the take-over of the feedforward path is shown in Fig. 5-7.

Due to additional parasitics in the levelshift circuits and due to influence of bonding wires the bandwidth of the circuit is set at 1 GHz. The basic topology of the three-stage 1-GHz amplifier which is depicted in Fig. 5-6 was already discussed. Transistor Q_{20} creates a common-mode feedback loop to fix the bias voltages of the intermediate stage Q_3, Q_4. A PTAT source consisting of transistors Q_{50}-Q_{55} and resistors R_{50}, R_{55} is used to bias the opamp [10]. To stabilize the PTAT source capacitor C_{51}, is added. Two versions

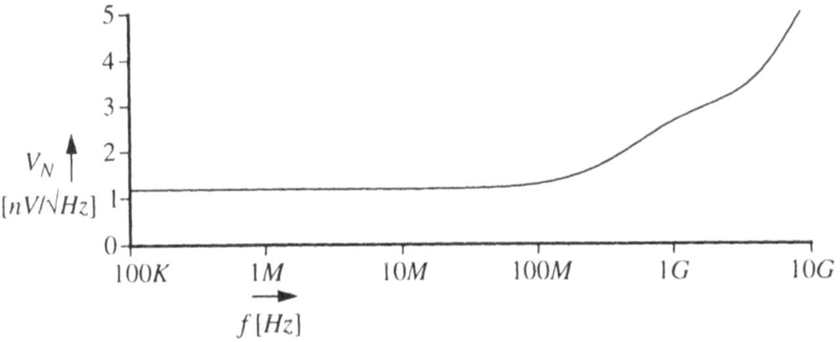

Fig. 5-7 *Typical input-referred noise characteristic of multipath-nested-Miller compensated wideband amplifier.*

have been made, an open-loop version, and a circuit with on-chip unity-gain feedback. A photomicrograph of the operational amplifier is shown in Fig. 5-8. The active die area is 0.26 mm^2.

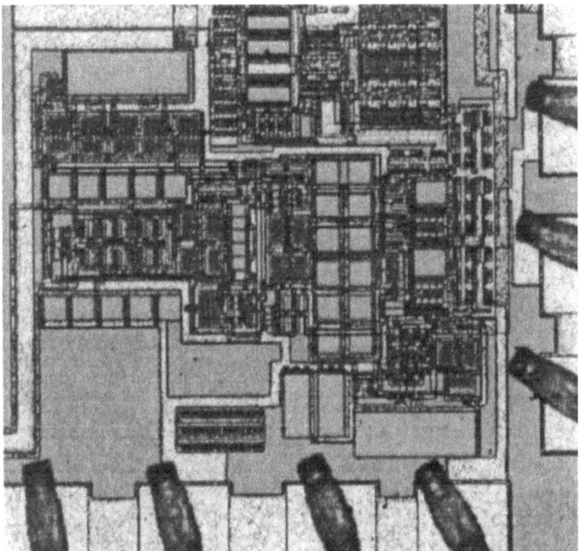

Fig. 5-8 *Photomicrograph of 1-GHz operational amplifier.*

Apart from circuit design, packaging is a very important issue at these high frequencies. The inductance of the bonding wires is a potential cause of oscillation problems. Each bonding wire loaded by capacitance forms a resonator. When we consider a small bonding wire with an inductance of 3 *nH* loaded by a capacitance of 1 *pF*, the resonant frequency is about 3 *GHz*. This is very close to the bandwidth of our circuit! Such a resonator within the feedback loop would be disastrous. Therefore, the capacitive loading of the load was separated from the feedback path by making two output pins, one for the load and one for the feedback. In addition, double bonding wires are used on the outputs and four bond wires on the supply lines. Double bonding wires are not necessary on the inputs, because the capacitive loading of the NPN input stage is very small. The pin connection scheme of the open-loop version is shown in Fig. 5-9a. The pin connection of the unity-gain connected opamp is shown in Fig. 5-9b. The chip is packaged in a small plastic SMD package. To test the circuit, the devices were mounted

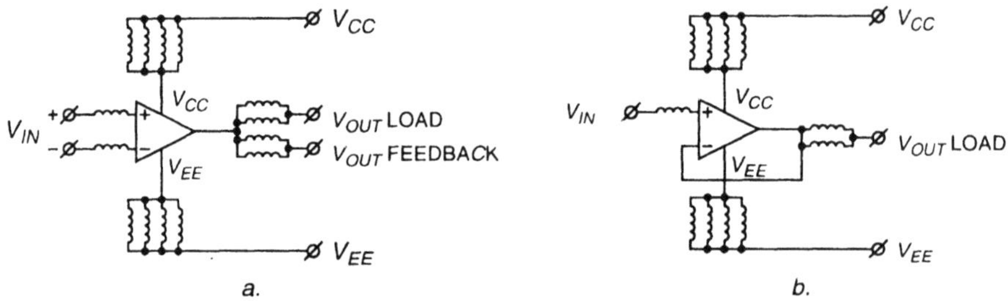

Fig. 5-9 *Pin connection scheme of 1-GHz operational amplifiers: open-loop amplifier (a) and unity-gain connected amplifier (b).*

on a copper plate together with small SMD external components and high-frequency SMA connectors connected as closely as possible to the pins. An HP8753A network analyzer was used to measure the frequency response. The measurement set-up is shown in Fig. 5-10. To compensate for delay in the cables, first a calibration measurement was performed as shown in Fig. 5-10a with the source cable directly connected to the load cable. Then the set-up as shown in Fig. 5-10b was used to measure the unity-gain connected device, and the set-up shown in Fig. 5-10c to measure the open-loop opamp. Decoupling capacitor *C* consists of several types of capacitors to cover the whole frequency range. Using this method an overview of the frequency response can be obtained, but it is not suited for determining the phase margin. In order to determine the phase margin, the voltage of the feedback output in Fig. 5-9a has to be examined. However, if a cable is connected to this node, the voltage is affected for high frequencies. Therefore, a separate measurement was performed to determine the phase margin, where

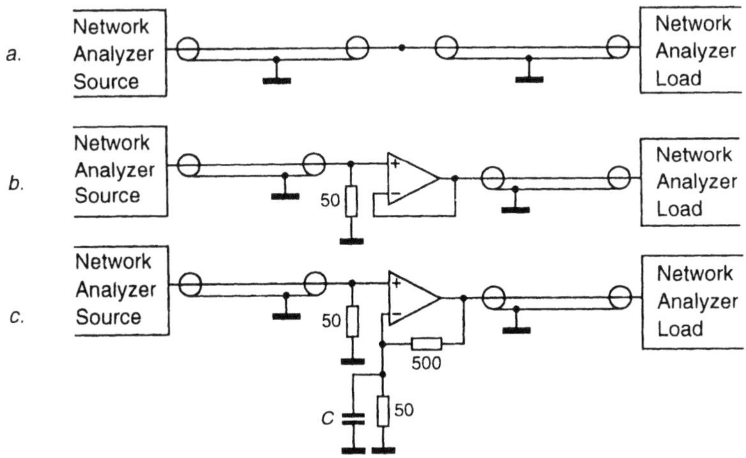

Fig. 5-10 *Measurement setup for 1-GHz operational amplifiers: calibration measurement (a), measurement of unity-gain amplifier (b) and open-loop amplifier (c).*

the output of the opamp was loaded by a 50 Ω resistor connected as close as possible to the output pins. The measured frequency response of 1-*GHz* operational amplifiers is shown in Fig. 5-11. The Bode plot of the unity-gain buffer is shown in Fig. 5-11a. The -3-*dB* frequency of the buffer is 1.5 *GHz*. Fig. 5-11b shows an overview of the frequency response from 300*KHz* to 3*GHz* of the open-loop version. The result of the more accurate measurement with resistive load to determine the phase margin is given in Fig. 5-12. The phase margin is 47°.

The step response of the on-chip unity-gain device was measured using a HP54120A 20GHz digitizing oscilloscope. This oscilloscope generates a 200mV step in a 50-Ω load. The response to the 200*mV* step is plotted in Fig. 5-13a. Again, the opamp was loaded by a 50-ohm resistor. The 1% settling time is 3.4 ns and the 0.1% settling time is 6.0 *ns*. By removing the 50-Ω resistor at the input of the opamp, a 400*mV* step is created in order to measure the slew rate. The result is shown in Fig. 5-13b. The harmonic distortion of a 2 *V* peak-to-peak output voltage was measured using the opamp in a 100x configuration. The THD as a function of frequency is plotted in Fig. 5-14.

The most important parameters of the 1-*GHz* operational amplifier are summarized in Table5-1.

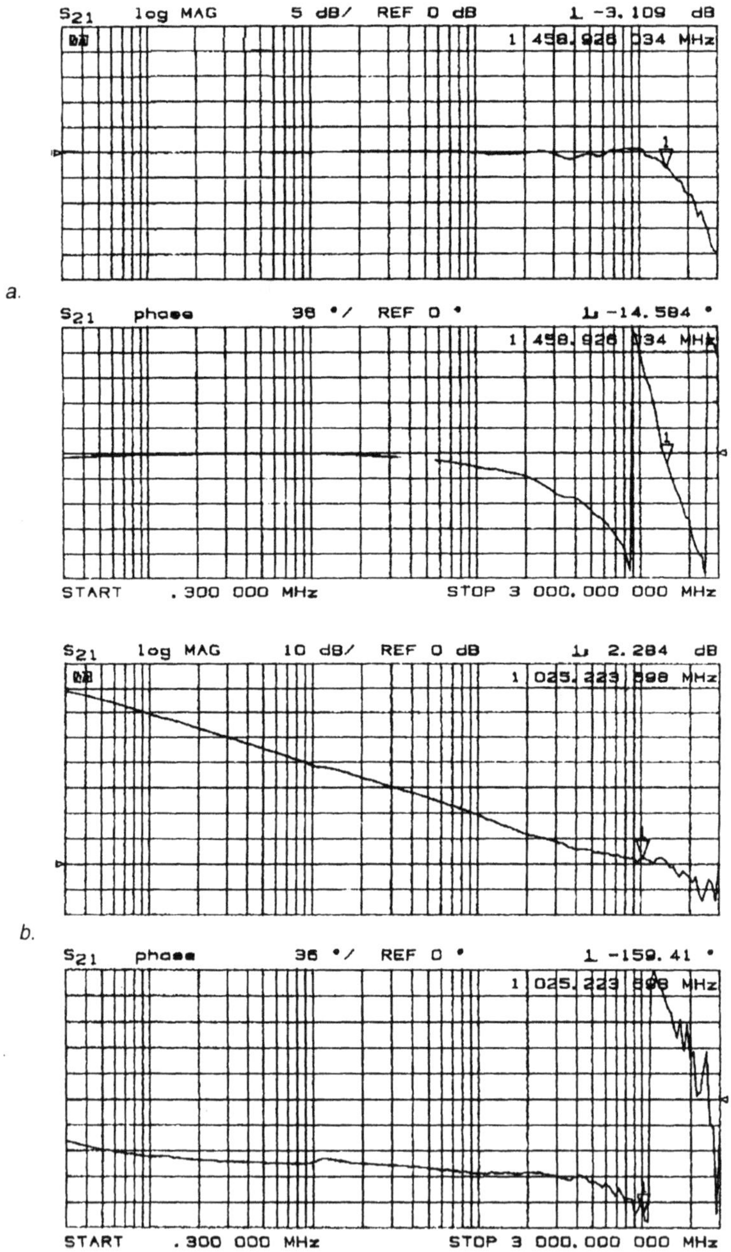

Fig. 5-11 *Bodeplots of 1-GHz operational amplifiers: on-chip unity-gain connected amplifier (a) and open-loop amplifier (b).*

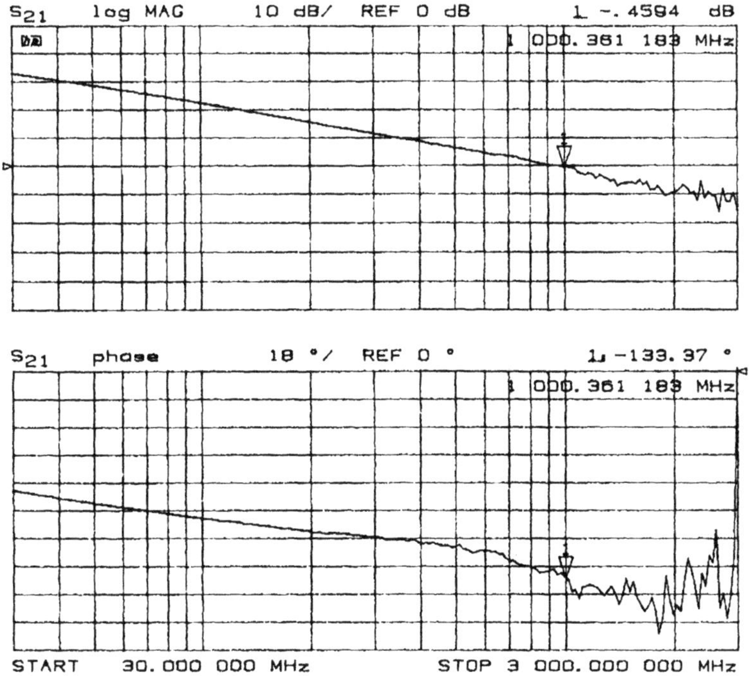

Fig. 5-12 *Measurement of frequency response for determining phase margin.*

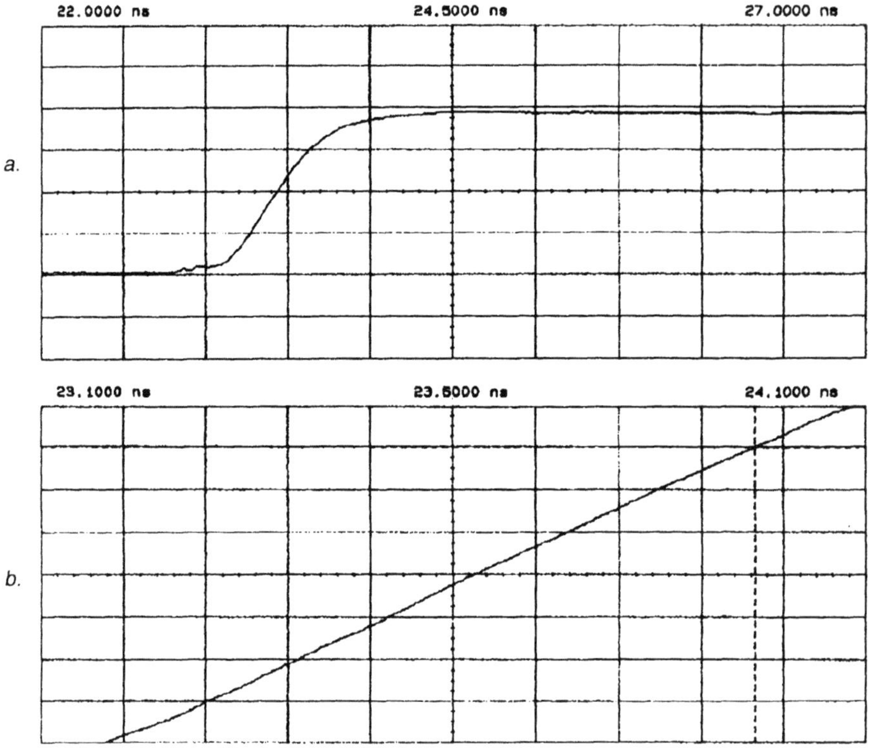

Fig. 5-13 *Step response of 1-GHz operational amplifier: response to a 200mV step (a) and response to a 400 mV step zoomed in to determine slew rate.*

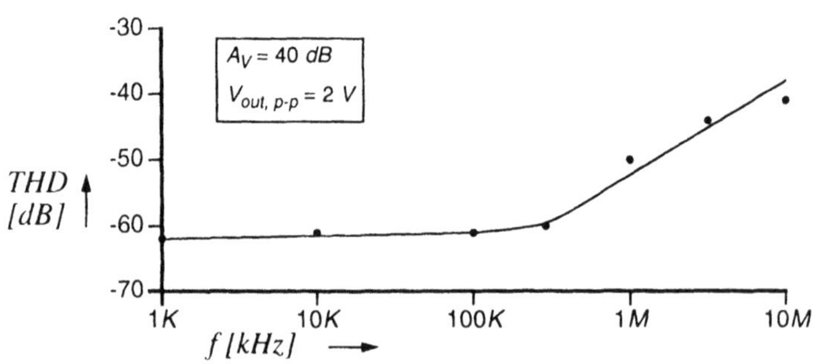

Fig. 5-14 *Measured THD as a function of frequency using 1-GHz opamp connected as 100x amplifier with 2V peak-to-peak output voltage*

Table 5-1 *Measured specifications of multipath nested Miller compensated 1-GHz operational amplifier (V_{PP}=5 V, R_L=50Ω, T_A=25 °C).*

symbol	parameter	value			unit
		min.	typ.	max.	
T_{op}	temperature range	-55		125	°C
V_{PP}	supply voltage range	3.6		5.5	V
I_{PP}	supply current		24		mA
I_B	input bias current			19	μA
V_{os}	offset voltage		0.2	0.5	mV
V_{CM}	common-mode input voltage range	1.0		V_{PP}-1.0	V
I_L	peak load current, sink and source	55	58	66	mA
A_{vol}	open-loop voltage gain		76		dB
V_{out}	output voltage swing	0.4		V_{PP}-1.2	V
SR	Slew rate		360		V/μs
$S_{1\%}$	1% Settling time (200mV step)		3.4		ns
$S_{0.1\%}$	0.1% Settling time (200mV step)		6.0		ns
BW	open-loop bandwidth		1		GHz
V_N	Equivalent input noise voltage		1.2		nV/√Hz

5.2.2 Power-efficient high-current amplifier

Instead of exploiting the bandwidth of the all-NPN output stage, the output-current capability can be exploited. This results in an amplifier with a high peak output current and very small stand-by current. This high current efficiency is combined with good voltage efficiency since the all-NPN output stage can deliver a peak-to-peak output voltage that is about a single base-emitter voltage smaller than the supply voltage. The basic circuit of the voltage and current efficient operational amplifier is depicted in Fig. 5-15. The circuit consists of all-NPN output stage Q_1, Q_2, intermediate stage Q_3, Q_4 and PNP input stage Q_5, Q_6. Since the resistive load is usually low in applications requiring high output currents, at least three stages are needed in order to obtain sufficient gain. The PNP input stage is applied in order to match the input-voltage range with the output-voltage range. For high current efficiency the intermediate stage and input stage are operated at small bias currents. Since the intermediate stage should deliver the base

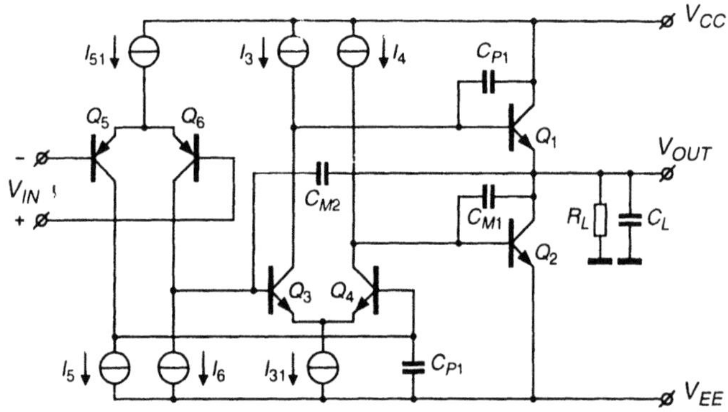

Fig. 5-15 *Simplified schematic of voltage and current efficient operational amplifier.*

currents of the output stage, the output transistors should be replaced by Darlington structures. The result would be a very large bias current in the output stage or a limited maximum output current. To obtain high efficiency the output transistors should therefore be replaced by a Darlington structure as shown in Fig. 5-16. The output

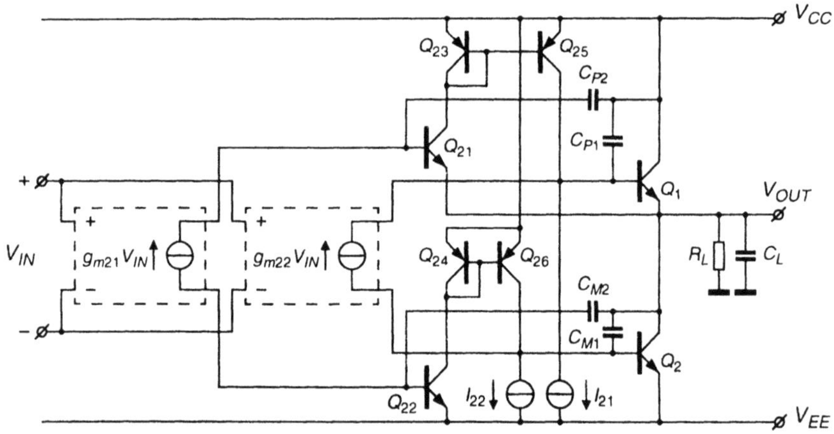

Fig. 5-16 *Voltage and current efficient output stage using multipath driven low-voltage Darlington circuits.*

transistors Q_1 and Q_2 are driven by driver transistors Q_{21} and Q_{22} via current mirrors Q_{23}, Q_{25} and Q_{24}, Q_{26}, respectively. The advantage of this configuration over conventional Darlington transistors is that no stacking of base-emitter voltages is used. Therefore the output voltage range is not further reduced and the output stage can still be used at relatively low supply voltages, depending on the maximum required output voltage. Frequency compensation is obtained using a nested-Miller structure consisting of capacitors C_{M1} and C_{M2} for the lower half of the output stage and capacitors C_{P1} and C_{P2} for the upper half. As with the wideband amplifier design, the two halves of the output stage are asymmetric but can be balanced by making capacitors C_{M2} and C_{P2} equal and by making C_{M1} and C_{P1} equal. The bandwidth reduction of nested Miller compensation is avoided by inserting a multipath [6,7,8] as indicated by g_{m22}. The complete circuit diagram of the voltage and current efficient amplifier is depicted in Fig. 5-17. The amplifier consists of output stage Q_1, Q_2, Q_{21}-Q_{26}, intermediate stage Q_3, Q_4, Q_{31}, Q_{41} and input stage Q_5, Q_6 with folded cascoded Q_{51}, Q_{61}. The folded cascode enables a high input voltage range that includes the negative supply rail V_{EE}. Transistor Q_{50} creates a common-mode feedback loop that fixes the voltage as the emitter of the intermediate-stage transistors. Intermediate stage transistors Q_{31} and Q_{41} implement a feedforward path that directly drives the output transistors Q_1, Q_2 and bypasses current-gain boosters Q_{21}, Q_{23}, Q_{25} and Q_{22}, Q_{24}, Q_{26}. Resistors R_{23} and R_{24} are used to counteract the reduction in current gain β_F at high output currents. Class-AB biasing of the output transistors is created using a feedback loop as discussed in chapter three. The bias current of the output transistors is measured by transistors Q_{11} and Q_{12}. The collector currents of Q_{11} and Q_{12} generate voltage drops across resistors R_{13} and R_{14}. The voltage drops across R_{13} and R_{14} drive decision pair Q_{15}, Q_{16} which forms the class-AB control amplifier together with Q_{17}. As output of the control amplifier the collectors of Q_{15} and Q_{16} are used that drive the inputs of the output stage via current mirror Q_{18}-Q_{20}. Thus a control loop is formed that control the minimum current of the output stage. By giving the current mirror two additional outputs using Q_{29} and Q_{30} that directly drive Q_1 and Q_2, a feedforward path is formed that guarantees the stability of the class-AB feedback loop. Diode-connected transistors Q_{13} and Q_{14} have been inserted to limit the voltage drop across resistors R_{13} and R_{14}, while resistors R_{11} and R_{12} are used to limit the current that flows in the measuring transistors at high output currents. Frequency compensation of the overall three-stage circuit is obtained using nested-Miller compensation consisting of inner capacitors C_{M2} and C_{P2} and outer capacitors C_{M3} and C_{P3}. Table summarizes the main specifications of the power-efficient operational amplifier. The interesting achievement of this amplifier is the very high ratio between peak output current and quiescent current.

This concludes the discussion on bipolar amplifiers. In the next section we move to a VLSI circuits where the minimum supply voltage and the size of the circuit are the dominant design parameters.

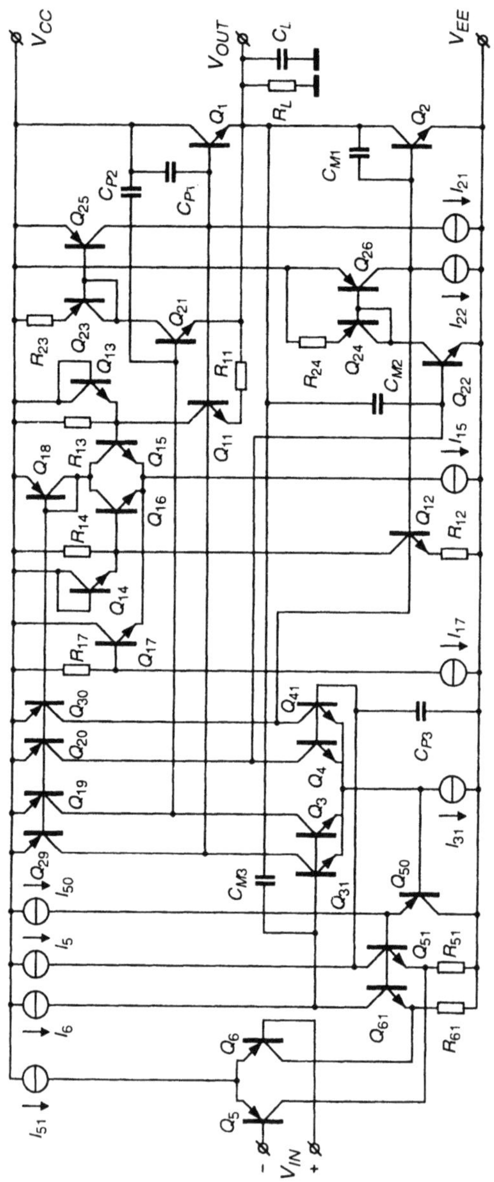

Fig. 5-17 *Complete circuit diagram of voltage and current efficient operational amplifier.*

Table 5-2 *Simulated specifications of power-efficient high-current operational amplifier (T_A=25 °C, R_L=1 kΩ, C_L=100 pF).*

Parameter	value	Unit
Temp. range	-55 .. 150	°C
Min. supply voltage	2.0	V
Supply current	350	μA
Peak output current	100	mA
Input offset voltage	0.5	mV
Input noise voltage (f=10 kHz)	10	nV/√Hz
CM input range	V_{EE}-0.5 .. V_{CC}-0.9	V
CM output-voltage range	V_{EE}+0.2 .. V_{CC}-0.2	V
CMRR	90	dB
Unity-gain frequency	1	MHz
Phase margin	65	°
dc-gain	80	dB

5.3 CMOS Compact amplifier topologies for VLSI

Applying simple topologies in order to obtain high efficiency is the key design technique for future analog cells in mixed-mode VLSI circuits. As the reduction in feature size of CMOS processes continues steadily the supply voltage has to be reduced while the higher integration density enforces lower power consumption per cell. Digital cells benefit from the size reduction yielding improved digital performance. Analog cells on the other hand, only benefit marginally, because minimum size transistors cannot be used due to noise and offset requirements. Further, the low supply voltage complicates the design often yielding more complex circuit solutions which may even result in a performance reduction. The only way for analog cells to keep up with the digital performance and the supply-voltage reduction is by using very efficient topologies that combine low-voltage operation with high power efficiency and small die area.

There are two main VLSI operational amplifier topologies. The first topology is the result of an attempt to minimize the die area yielding a compact two stage circuit as discussed in section 5.3.1. A drawback of the compact VLSI topology is that the lowest possible supply voltage cannot be reached. This lowest voltage can only be reached using at least three stages yielding the ultimate-low-voltage VLSI topology which is presented in section 5.3.3.

5.3.1 Compact 1.4*V* topology with rail-to-rail output range

The basic low-voltage compact opamp is shown in Fig. 5-18a. The amplifier consists of a PMOS input stage M_3, M_4, a current mirror M_7, M_8 with cascodes M_5, M_6 and, a rail-to-rail output stage M_1, M_2. A PMOS input stage is used to allow common-mode voltage down to and below the negative supply rail. The current mirror is needed to sum the

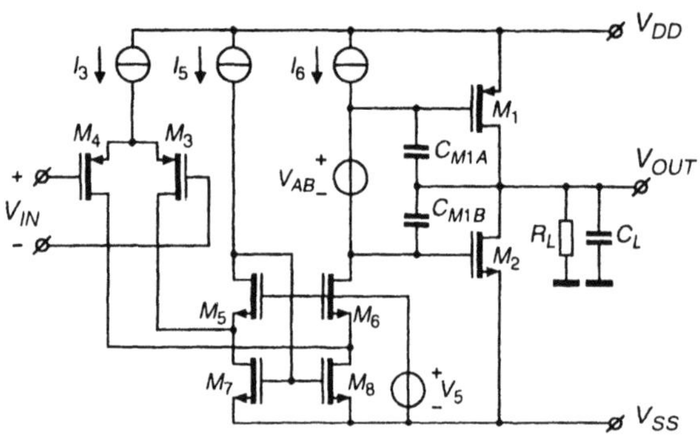

Fig. 5-18 *Compact low-voltage operational amplifier principles.*

opposite-phase signals of the differential input stage in order to drive the gates of the rail-to-rail output stage in phase. The cascodes provide the necessary levelshift between input and output stage. Further, M_6 provides gain by leaving the high input impedance of the gates of the output stage intact. The rail-to-rail output stage allows rail-to-rail output-signal swing making efficient use of the supply voltage. By biasing the output stage in class AB the supply current is used efficiently. The class-AB biasing is in principle represented by the voltage source V_{AB} which expresses all its important properties. In order to set the quiescent current the sum of the gate-source voltages of the output stage can be controlled in such a way that it is equal to the sum of a reference PMOS gate-source voltage $V_{GS,\,P}$ and an NMOS gate-source voltage $V_{GS,\,N}$, which is obtained by giving V_{AB} the value [11,12]

$$V_{AB} = V_{DD} - V_{SS} - V_{GS,\,P} - V_{GS,\,N} \tag{5-16}$$

Note that V_{AB} can be negative at low supply voltages and depending on the type of class-AB behavior V_{AB} may be signal dependent. Sufficient quiescent current is important to guarantee stability when the output voltage crosses the common-mode output voltage and no signal current is drawn by the load. Further, a smooth switch over from one transistor to the other is obtained avoiding gross cross-over distortion, because the output transistors are gradually switched on and off. A final property is that the signal from cascode M_6 is distributed over the two output transistors in such a way that all the signal contributes to the output voltage independent of which transistor is active. By biasing the class-AB control circuit by the current mirror, no extra current is needed and no extra noise and offset is generated by adding the class-AB function. Because of the simple two-stage topology Miller compensation is sufficient to compensate the opamp as indicated by Miller capacitors C_{M1A} and C_{M1B} which are connected across each output transistor. Thus, no difficult compensation scheme is needed that would require large bandwidth reduction or more bias current in the output stage.

In general, the minimum supply voltage is limited by the output stage, because it needs to drive high output currents requiring large gate-source voltages. The supply voltage of the output stage is equal to the sum of the gate-source voltage of M_1, the voltage V_{AB} and, the gate-source voltage of M_2 as can be concluded from Fig. 5-18. Since V_{AB} can be negative, theoretically the minimum supply voltage needed by the output stage can be very small. Practically however, the gate of an output transistor is always at least connected via a current source to the supply rails. Because we need to obtain sufficient gain in only two stages, the current source should not deteriorate the impedance and must be cascoded. To allow the lowest possible supply voltage the transistors of the current source should be biased at the edge of the triode region resulting in a minimum supply voltage limited by a gate-source voltage and two saturation voltages.

Concluding, each component of the compact amplifiers as shown in Fig. 5-18 has at least one important function proving the high efficiency in terms of the number of components. Because of the low number of components and the two stage topology which can be efficiently compensated using Miller compensation, a relatively small bias current is required.

An existing realization of the principle shown in Fig. 5-18 is presented in Fig. 5-19. This circuit is used in conventional compact opamps [13,14,15]. The class-AB voltage source V_{AB} is implemented using two complementary head-to-tail connected transistors M_{11} and M_{12} together with biasing transistors M_{21}-M_{24} [16] as described in section 4.4. The biasing of the output transistors M_1 and M_2 is strongly controlled by two MOS translinear loops M_1,M_{11}, M_{23}, M_{21} and M_2, M_{12}, M_{24}, M_{22}, respectively, fixing a well defined quiescent current. Further, cutoff is prevented, because if one output transistor is driven hard, the gate-source voltage of the other output transistor is clamped by one of the control transistors M_{11} or M_{12} leaving a certain minimum bias current in the inactive

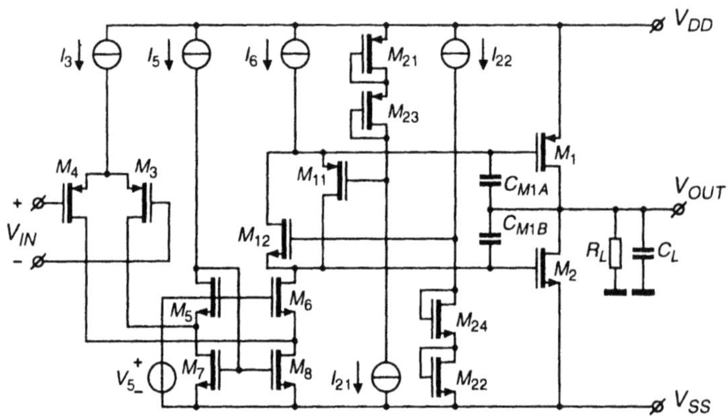

Fig. 5-19 *Conventional compact two-stage operational amplifier with rail-to-rail feedforward class-AB output stage with mesh.*

output transistor. As discussed in chapter three, the class-AB control circuit couples the gates of the output transistors in the quiescent mode of operation by generating a low impedance. When one output transistor delivers a large output current the other output transistor is regulated at a constant drain current and the gates of the output transistors are decoupling by generating a large impedance. At the same time, however, the class-AB control circuit steers all the signal to the active output transistor. Another important property of the class-AB control circuit is the contribution to the noise and offset. Fortunately, the class-AB control transistors M_{11} and M_{12} are biased by using the bias current which is already needed to bias the summing circuit M_5-M_8. Therefore, no additional bias-current sources are required that would increase the noise and offset and that would also reduce the gain [13]. The remaining disadvantage of the circuit is that the minimum supply voltage is equal to two gate-source voltages and one saturation voltage. For example, we need to bias the gate-source voltages of M_1 and M_{11} and sufficient voltage across current source I_{21} inside the supply voltage.

To allow lower supply voltages the voltages at the gates of M_{11} and M_{12} should be one gate-source voltage closer to the supply rails, which can be obtained by removing transistors M_{23} and M_{24} and placing these transistors as a levelshift in the mesh. This results in a first step towards a low-voltage compact amplifier as shown in Fig. 5-20. The mesh now consists of transistors M_{11} and M_{12} and diode-connected transistors M_{23} and M_{24}. The output transistors are biased by MOS translinear loops M_1, M_{23}, M_{11}, M_{21} and M_2, M_{24}, M_{12}, M_{22}. Some of the properties of the previous circuit are maintained. However, the diode levelshifts introduce two problems. First, cutoff of the output transistors occurs when the output stage is driven hard. If for example M_1 delivers a high output current its gate-source voltage is large and all the current in the mesh flows

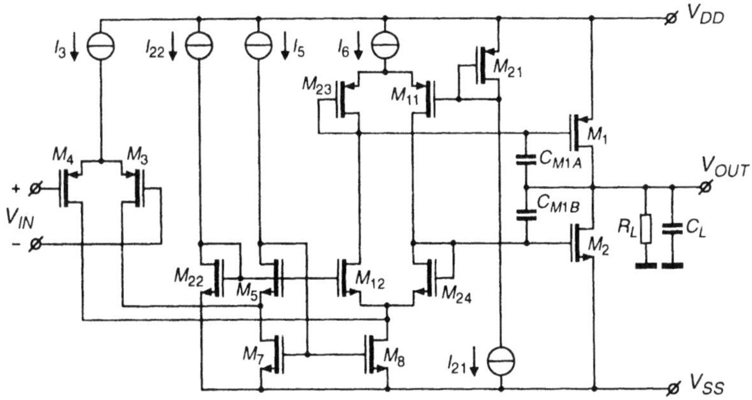

Fig. 5-20 *First step to a low-voltage compact two-stage operational amplifier with rail-to-rail feedforward class-AB output stage using level-shift in mesh.*

through M_{23} and M_{12} while no current flows through M_{11} and M_{24}. The voltage across M_{24} is very small and therefore the gate-source voltage of M_2 is small and cutoff occurs. A second problem is that there is almost no room to bias the current source I_6 and transistor M_8. Especially when we note that the output transistors usually have large W/L ratios in order to drive large output currents. Therefore in the quiescent state the gate-source voltage of the output transistors is relatively small and room for M_8 and I_6 can only be created by making the W/L ratio of M_{11}, M_{12}, M_{23} and M_{24} very large. There is certainly no room to insert cascodes. Due to the diode connection M_{23} and M_{24} also do not function as a cascode so that the obtainable gain is too low in a two-stage topology.

The problems discussed above are related to the diode connection of M_{23} and M_{24}. If we can remove this diode connection, we can make transistors M_{23} and M_{24} operate as cascodes and room for M_8 and I_6 can be created more easily. The principle of the resulting low-voltage compact topology is presented in Fig. 5-21. The mesh consisting of transistors M_6, M_{16}, M_9, M_{19} is folded and biased independent of the output transistors. In order to obtain the desired class-AB biasing a feedback loop is created consisting of measuring transistors M_{11} and M_{12}, and a class-AB control block containing a circuit that implements the desired class-AB control law and generates a control voltage V_{AB} to drive folded-mesh transistors M_6 and M_{16}. The current of output transistor M_1 is measured by M_{11} and is fed into the control block. The current of output transistor M_2 is measured by M_{12} and is also fed into the control block. Transistors M_6 and M_{16} form a control amplifier that regulates the signal at the inputs of the output transistors in such a way that the control voltage is equal to a reference voltage formed by M_{22}. In this way a

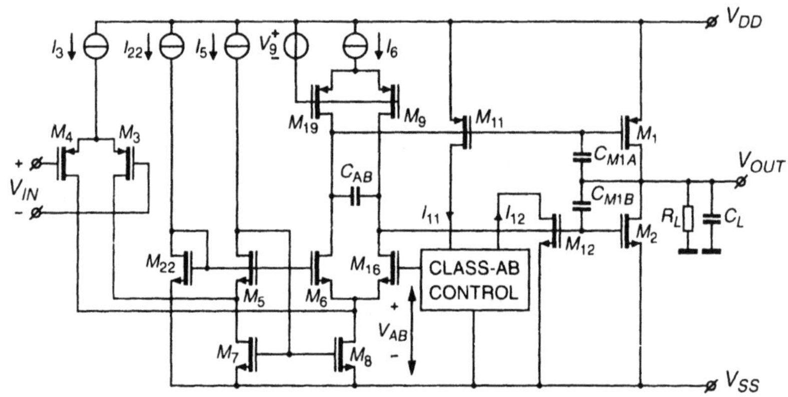

Fig. 5-21 *Low-voltage compact two-stage operational amplifier with rail-to-rail feedback class-AB output stage with folded mesh and class-AB control.*

feedback loop is created which enforces the desired class-AB behavior. Voltage source V_9 forms a fixed bias voltage for transistors M_9 and M_{19}. Preferably, the class-AB control block contains a circuit that not only controls the quiescent current of the output stage, but also maintains a minimum current. Such a circuit is usually called a minimum selector circuit.

The feedback class-AB control circuit with folded mesh has all the desired properties. The minimum supply voltage is one gate-source voltage and two saturation voltages. We need for example one gate-source voltage for M_1, a saturation voltage to bias M_{14} and another saturation voltage to bias M_{10}. Further, the folded mesh M_{13}-M_{16} distributes the signal from the input stage over both output transistors and, because the mesh is only connected to high impedances the signal cannot be lost and all the signal is used to drive the output transistors. This is also the reason why the mesh transistors function as cascodes. Normally, when we tie the sources of two cascodes together, the output impedances are degraded. If we consider for exaiaple M_{14}, M_{16} as a differential pair and connect the gate of M_1 to a low impedance, the output impedance at the gate of M_2 would be of the order of the output impedance r_d of the transistors. But because the mesh is only connected to high impedances M_{16} functions as a cascode and the impedance at the gate of M_2 is of the order of r_d of M_{10} multiplied by the voltage gain μ of M_{16}. The mesh still generates a low impedance between the gates of the output transistors as shown in section 3.5.

Again, the contribution of the mesh transistors to the noise and offset should be small. In the same way as in the conventional circuit the mesh transistors use the bias current of the summing circuit. Therefore, no additional bias current is required and no extra noise or offset is generated. The differential pair M_{14}, M_{16} may give reason to concern since differential-pair transistors normally contribute to the noise. However, these transistors operate only as differential pair for class-AB signals which do not contribute to the output signal. For normal signals the mesh transistors operate as cascodes which do not contribute to the noise.

Because a feedback loop is used to implement the class-AB control, the stability of the class-AB loop should also be considered. The loop consists of class-AB amplifier M_{14}, M_{16}, measuring transistors M_{11} and M_{12}, and the class-AB control block. Basically, the class-AB loop can be considered as a single stage amplifier with transconductance stage M_{14}, M_{16} and with the Miller capacitors and the gate-source capacitors as load capacitors as discussed in section 3.5. The bandwidth and phase margin of the class-AB loop can be manipulated by adjusting the transconductance of the class-AB amplifier M_{14}, M_{16} or by inserting compensation capacitor C_{AB} as shown in Fig. 5-21. In chapter three more details about the operation of the class-AB loop and the frequency compensation are discussed.

We now only have to select a suitable implementation of the class-AB block. The simple minimum-selector circuit [17] is well suited to be applied with the folded mesh control circuit. The complete compact operational amplifier with class-AB output stage that incorporates this minimum selector in a folded mesh feedback circuit is shown in Fig. 5-22 [18,19]. The opamp has a PMOS input stage M_3, M_4 allowing common-mode input voltages down to the negative rail. The input stage drives a cascoding and summing circuit consisting of a current mirror M_7, M_8 and cascodes M_5, M_6, M_{15}, M_{16}, M_9, M_{19}. The summing circuit drives the output stage M_1, M_2 with class-AB control M_{10}-M_{14}. The class-AB control uses the simple minimum selector M_{11}, M_{13}, M_{14} as discussed in chapter 4. The minimum selector drives the class-AB amplifier M_6, M_{16} using transistor M_{10}. The class-AB amplifier is part of the folded mesh M_6, M_{16}, M_9, M_{19}. The current of M_2 is measured by M_{12}. The drain current of M_{12} flows through transistor M_{14} which is part of the minimum selector circuit M_{11}, M_{13}, M_{14}. A good class-AB behavior is obtained using an absolute minimum of components and only very little current since we only need to bias the two branches needed to measure the currents in the output transistors. Also, this minimum selector circuit can operate on the lowest supply voltage possible since only one gate-source voltage and one saturation voltage is needed. Of course, in combination with the folded mesh an additional saturation voltage is required. Detailed operation of the class-AB control circuit as well as a plot of the class-AB characteristic is presented in section 3.5.

A modification shown in Fig. 5-22 is that cascode M_5 is splitted into two parts, M_5 and M_{15} of which M_{15} is driven by the class-AB control loop. The reason for this is that the signals in the class-AB loop are present across the finite output impedance of M_8.

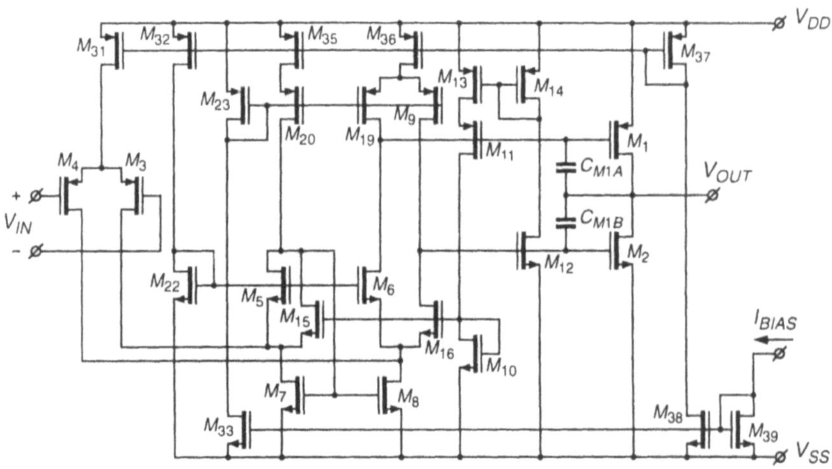

Fig. 5-22 *Complete circuit diagram of 1.7-V compact operational amplifier with PMOS input stage and folded mesh with simple minimum selector.*

Therefore, the drain current of M_8 is influenced by the class-AB loop which interferes with the signal of the main amplifier. The result is that the dc-gain of the amplifier is modulated by the output current. The dc-gain may even change polarity! The influence of the class-AB loop on the dc gain can simply be cancelled using M_{15} which drives the same signals across the output impedance of M_7. These signals flow through current mirror M_7, M_8 and cancel the signal generated by M_{16}. The amplifier has been realized in a 1.6-μm CMOS process. A micrograph of the circuit is depicted in Fig. 5-23. The

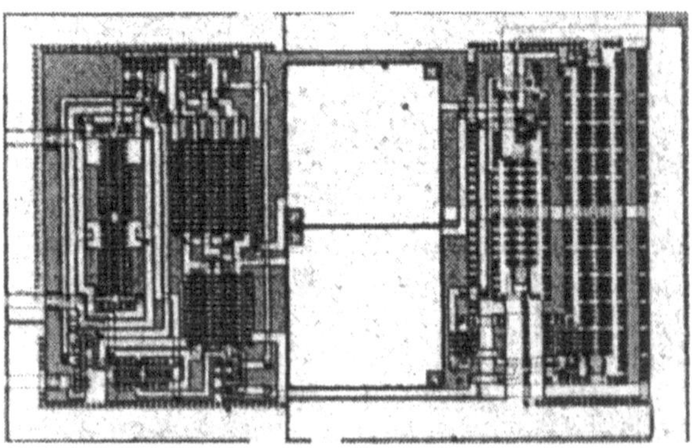

Fig. 5-23 *Photomicrograph of 1.7-V compact operational amplifier.*

amplifier occupies a die area of 0.08 mm^2. Due to the relatively high PMOS threshold voltage of 1.1 V, the minimum supply voltage of 1.7 V is relatively high. The frequency response is plotted in Fig. 5-24 demonstrating a 4-*MHz* unity-gain frequency with 62° phase margin at a load capacitance of 5 *pF*. The dc-gain is 86 *dB* at a resistive load of 10

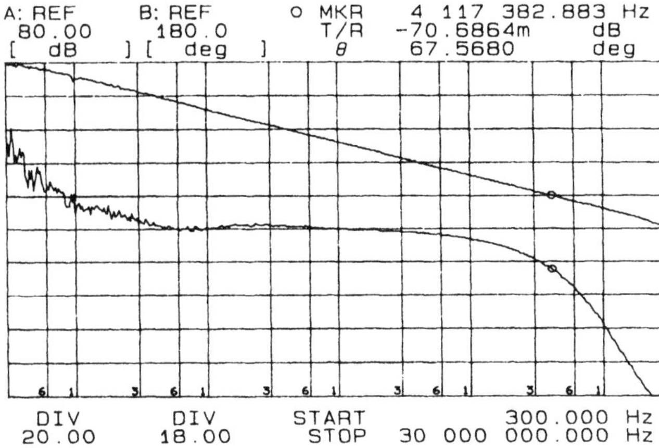

Fig. 5-24 *Frequency response of 1.7-V compact operational amplifier.*

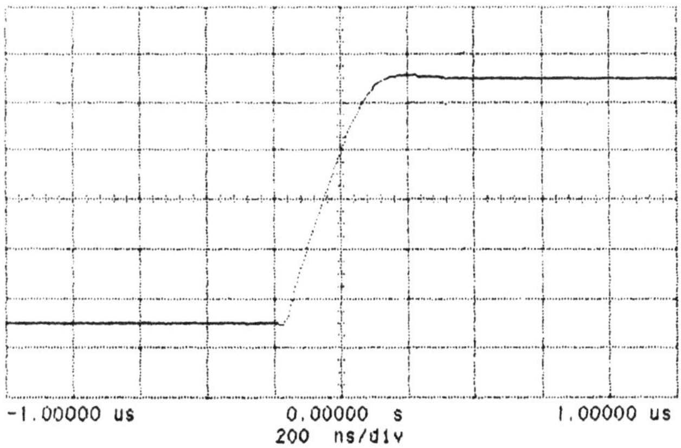

Fig. 5-25 *Response to a 1-V step of 1.7-V compact operational amplifier (y-axis scale 200 mV/div).*

*k*Ω. The response to a 1*V* step is shown in Fig. 5-25. The amplifier settles to 1% within 380 *ns* and to 0.1% within 465 *ns*.

Alternatively, a PMOS current mirror can be used in the summing circuit. A compact operational amplifier with PMOS current mirror in the summing circuit is shown in Fig. 5-26. Compared to the circuit depicted in Fig. 5-22 the NMOS current mirror is replaced

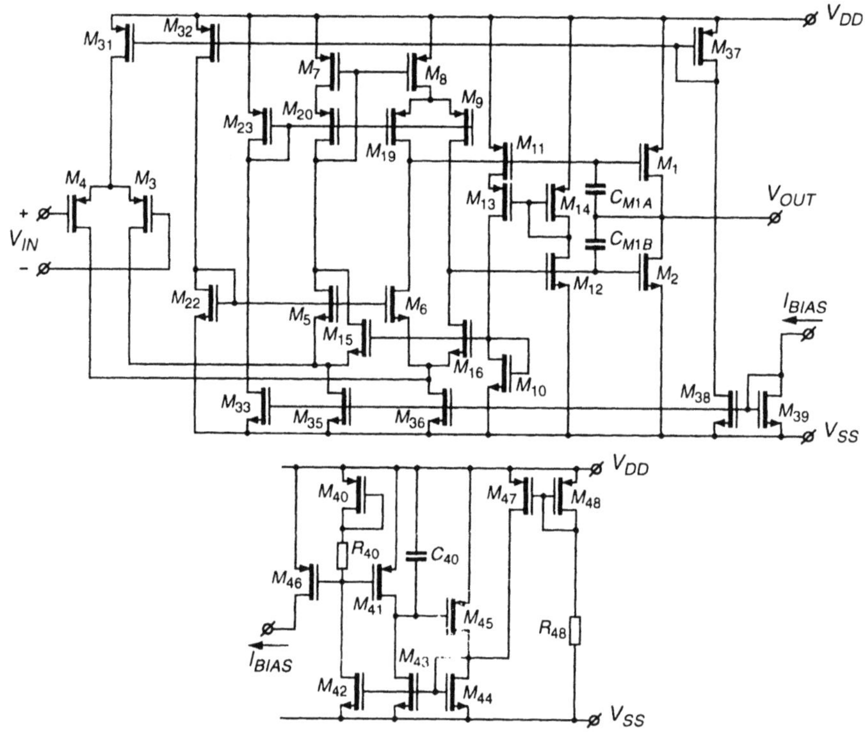

Fig. 5-26 *Complete circuit diagram of 1.4-V compact two-stage operational amplifier with PMOS current mirror in summing circuit and PTIM bias-current source.*

by current source I_5 and I_6, while the current sources are replaced by PMOS current mirror M_7, M_8. The advantage of the PMOS current mirror is that less bias current is required. In both cases the current sources I_5 and I_6 are equal to two times I_3. Using the PMOS current mirror the total bias current of input stage and cascodes is equal to the sum of I_5 and I_6 while using the NMOS current mirror the total bias current is equal to the sum of I_3, I_5 and I_6. Thus, 50% more bias current is needed when the NMOS current mirror is applied. However, the bandwidth of the PMOS current mirror is smaller. Especially, if the amplifier is designed for low capacitive loads resulting in relatively high bandwidths, the application of a PMOS current mirror in the summing circuit causes problems. The compact amplifier with PMOS current mirror has been implemented using the CMOS part of a 0.8μm BiCMOS process. The bias current source as depicted in Fig. 5-26 has been applied to generate the bias current of the amplifier so that the bandwidth is nearly independent of temperature. The current source is a CMOS version of a low-voltage PTAT current source discussed in chapter two. PMOS transistors are used to generate the current in order to obtain good matching with the PMOS input transistors. A micrograph of the amplfier is shown in Fig. 5-27. The

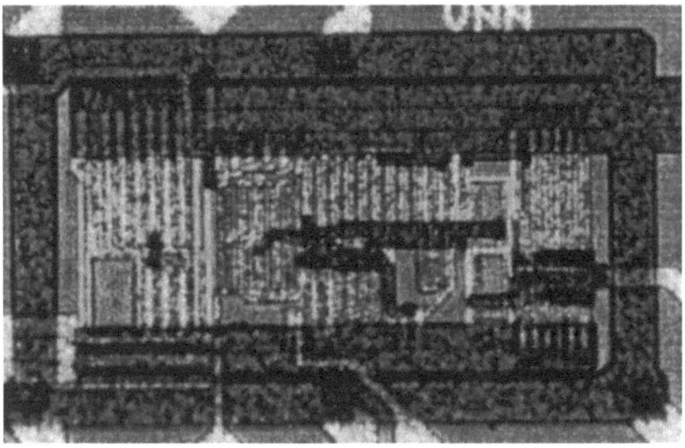

Fig. 5-27 *Photomicrograph of 1.4 V compact operational amplifier.*

frequency response of the amplifier is plotted in Fig. 5-28. It achieves a unity-gain frequency of 14 *MHz* with a phase margin of 50°. The minimum supply voltage is 1.4 *V* using a process with PMOS threshold voltage of 0.8 *V*. The response to a 1 *V* step is shown in Fig. 5-28. The amplifier settles to 1% within 150 *ns* and to 0.1% within 360 *ns*. Table5-2 summarizes the measured specifications of the two compact operational amplifiers.

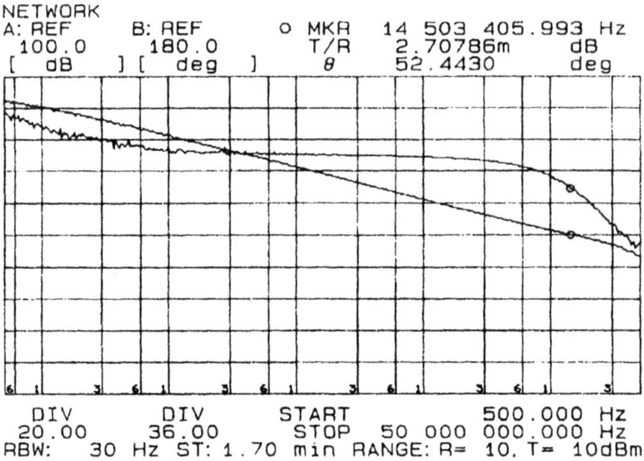

Fig. 5-28 *Frequency response of 1.4-V compact amplifier.*

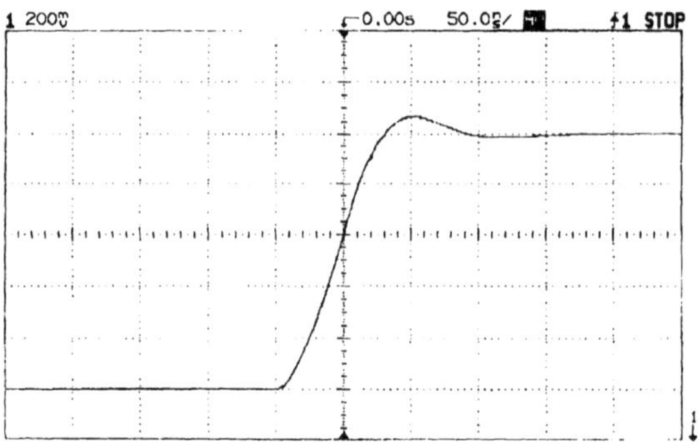

Fig. 5-29 *Response to a 1-V step of 1.4-V compact amplifier,*

Table 5-3 *Measured specifications of compact operational amplifiers (T_A=25 °C, R_L=10 kΩ, C_L=5 pF).*

Parameter	1.7-*V* compact amplifier (V_{DD} = 1.8 *V*)	1.4-*V* compact amplifier (V_{DD} = 1.5 *V*)	Unit
Die area	0.08	0.026	mm^2
Temp. range	-55 .. 150	-55 .. 150	°C
Min. supply voltage	1.7	1.4	V
Supply current	180	280	μA
Peak output current	1.2	2.4	mA
Input offset voltage	5	5	mV
Input noise voltage (*f*=100 *kHz*)	33	50	nV/√Hz
CM input range	V_{SS}-0.5 .. V_{DD}-1.3	V_{SS}-0.5 .. V_{DD}-1.2	V
CM output-voltage range	V_{SS}+0.2 .. V_{DD}-0.2	V_{SS}+0.2 .. V_{DD}-0.2	V
CMRR	80	80	dB
Unity-gain frequency	4	14	MHz
Phase margin	67	50	°
dc-gain	86	86	dB
Slew rate	4	19	V/μs
Settling time (1%, 1*V*)	380	150	ns
Settling time (0.1%, 1*V*)	480	360	ns

5.3.2 Compact 1.7*V* topology with rail-to-rail input and output ranges

The low-voltage compact opamp topology can be extended to a circuit with a rail-to-rail input stage. For full rail-to-rail operation, a rail-to-rail input stage requires a supply voltage of two gate-source voltages and two saturation voltages which is higher than the minimum supply voltage of the basic low-voltage compact opamp. Because the currents handled by the input stage are relatively small, the minimum supply voltage can be lower than the supply voltage required by conventional output stages. Therefore it is interesting to consider a low-voltage compact opamp with rail-to-rail input stage.

Main problem of the rail-to-rail input stage is that the output currents of the rail-to-rail input stage vary as a function of the common-mode voltage. In order to process these signals without affecting the bias control of the output stage a special structure of the summing circuit is required as discussed in chapter 3 using a floating current source. The problem is not as serious as when a feedforward biased output stage is used because then the biasing of the output-stage transistors is affected. In this case only the biasing of the class-AB amplifier is changed. This should still be avoided as much as possible since it results in a variation of the bandwidth of the class-AB control loop. In the compact operational amplifier with rail-to-rail input stage as shown in Fig. 5-30 a second mesh is used to implement the floating current source. The circuit consists of a PMOS input pair

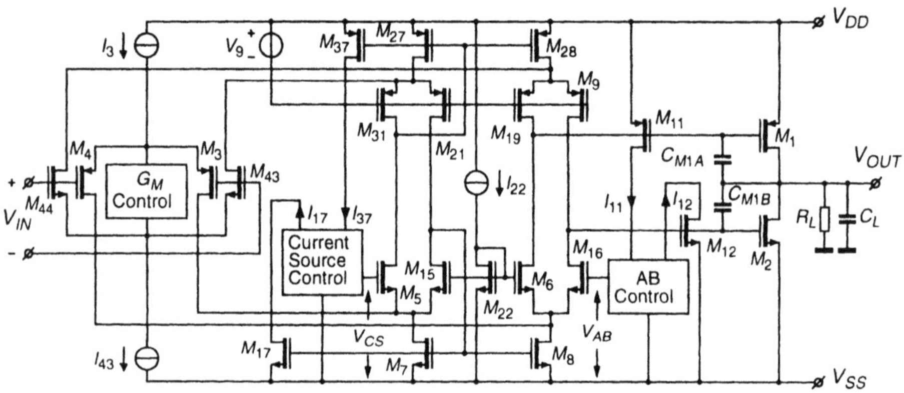

Fig. 5-30 *Low-voltage compact operational amplifier principle with rail-to-rail input stage.*

M_3, M_4, an NMOS input pair M_{43}, M_{44} a summing circuit M_5-M_9, M_{15}, M_{16}, M_{19}, M_{21}, M_{27}, M_{28}, M_{31} and, an output stage M_1, M_2. A class-AB control circuit and a g_m-control circuit still have to be inserted and are indicated by the blocks. A first mesh M_6, M_{16}, M_9, M_{19} is used to implement the class-AB control by controlling the gates of the output transistors M_1, M_2. The summing circuit consists of two current mirrors M_7, M_8 and M_{27}, M_{28} and a floating current source. A second mesh M_5, M_{15}, M_{21}, M_{31} is used to create the floating current source by controlling the gates of the mirror transistors M_7, M_8 and M_{27}, M_{28}. The biasing of these mirrors is controlled in the same way as with the class-AB control by means of a feedback loop. The bias current of the PMOS mirror M_{27}, M_{28} is measured using M_{37}. The biasing of the NMOS current mirror M_7, M_8 is measured by M_{17}. The drain currents of M_{17} and M_{37} flow into the current-source control block which completes the feedback loop by driving the gate of M_5. The current-source control can be implement by using the same circuits as used for controlling the output stage.

An example of a complete compact operational amplifier with rail-to-rail input stage is shown in Fig. 5-31. The g_m control of the input stage is created using current switches M_{45}, M_{46} [12,14] as discussed in chapter 3. Part of the bias current of the PMOS pair M_3,

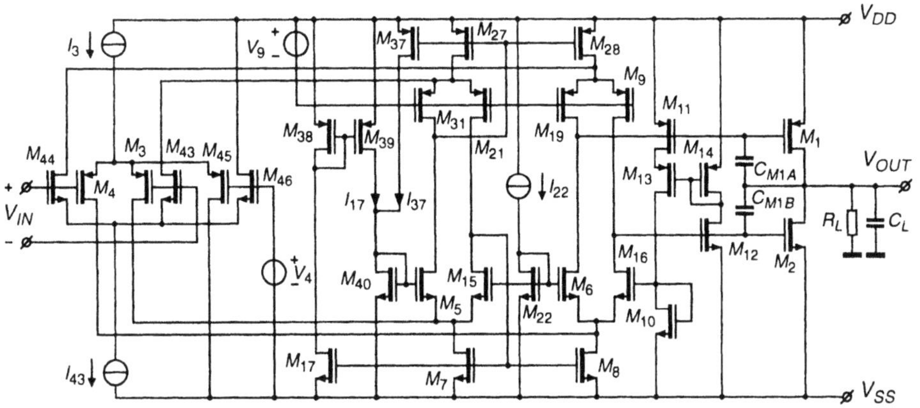

Fig. 5-31 *Implementation of compact amplifier with rail-to-rail input stage.*

M_4 is removed in the middle of the common-mode input-voltage range by the current switch M_{45}. At the same time, switch M_{46} removes part of the bias current of the NMOS input pair M_{43}, M_{44}. For high common-mode input voltages the NMOS input pair operates and the PMOS input pair is off, while the bias current of the PMOS pair completely flows through the switch M_{45}. For low common-mode input voltages, the PMOS input stage functions and all the bias current of the NMOS pair flows through the switch M_{46} so that the NMOS input stage is off. In the middle of the common mode range, when the common-mode voltage is equal to the reference voltage V_4 both input stages are operating together and therefore the g_m of each input stage should be a factor two smaller than when only one input stage functions. Consequently, the current through the input stage transistors should be a factor 4 smaller, which can be obtained by assuring that 3/4 of the bias current flows through the current switches. This is obtained if the W/L ratio of the current switches is a factor 3 larger than the W/L ratio of the input transistors.

The biasing of the mesh M_5, M_{15}, M_{21}, M_{31} which implements the floating current source is obtained using a class-A control consisting of measuring transistors M_{17}, M_{37} and current mirror M_{39}, M_{39}. A class-A control is sufficient because the bias current in the summing circuit is relatively small. The bias current of the PMOS mirror M_{27}, M_{28} is measured using M_{37}. The bias current of the NMOS current mirror M_7, M_8 is measured by M_{17}.The drain current of M_{17} flows through mirror M_{38}, M_{39} and is added to the drain current of M_{37}. The sum of these current flows through M_{40} which steers the control amplifier M_5, M_{15} that regulates the current through the current mirrors M_7, M_8 and M_{27}, M_{28} in such a way that their sum is equal to a reference current I_{22} which flows through M_{22}. Thus the sum of the bias current of the PMOS mirror and the NMOS mirror is regulated constant yielding a class-A biasing.

An alternative way of realizing a compact amplifier with rail-to-rail input stage is depicted in Fig. 5-32. The g_m-control is implemented using switches M_{45}-M_{48} which add the control current to the outputs of the inputs stages. By using this configuration the output currents of the inputs stage do not vary as a function of the common-mode voltage as discussed in chapter 3. Therefore, the same simple summing circuit can be applied as used in the previous compact amplifiers with single PMOS input stage. A disadvantage is that the current-switch transistors contribute to the noise. The contribution to the noise can be reduced by giving the switching transistors a small *W/L* ratio. Because, increasing the *W/L* ratio leads to a larger gate-source voltage, a levelshift voltage must be inserted between the gates of the PMOS and the NMOS switches. In this example the levelshift is created by using to bias voltages using M_{55} and M_{56} where the voltage created by M_{56} is larger by using a smaller *W/L* ratio. Another advantage is that due to the smaller switches

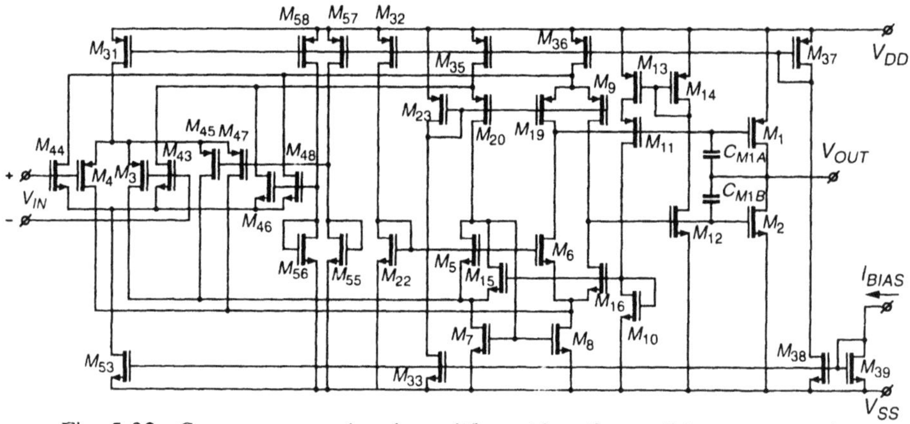

Fig. 5-32 *Compact operational amplifier with rail-to-rail input stage using current switches with constant common-mode output current.*

the die area is reduced. The amplifier has been implemented in a 1.6-μm CMOS process. A micrograph of the circuit is depicted in Fig. 5-33. The frequency response is plotted in Fig. 5-34. The amplifier has a unity-gain frequency of 4 *MHz* and a phase margin of 62° using a 5-*pF* load. The response to a 1V step is presented in Fig. 5-35. The amplifier has a 1% settling time of 370 *ns*, a 0.1% settling time of 465 *ns* and a slew rate of 4.5 *V/μs*. The offset and transconductance as a function of the common-mode input voltage are

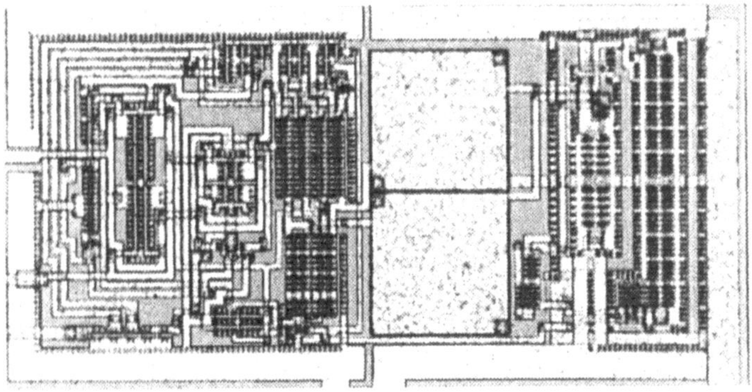

Fig. 5-33 *Micrograph of compact operational amplifier with rail-to-rail input stage.*

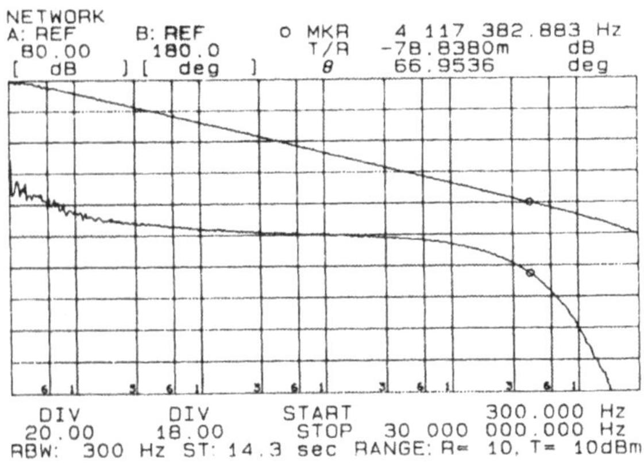

Fig. 5-34 *Frequency response of compact operational amplifier with rail-to-rail input stage.*

shown in Fig. 5-36a and Fig. 5-36b, respectively. The offset voltage changes as a function of the common-mode input voltage from the offset of the PMOS input stage for low values of the common-mode input voltage to the offset of the NMOS input stage for high values of the common-mode input voltage. The offset varies approximately 8 mV which decreases the CMMR to about 40 *dB* for signals around the switch-over region of the g_m-control current switches. The transconductance plot is based on measurement of the unity-gain frequency. From the plot it can be concluded that the g_m and the unity-gain

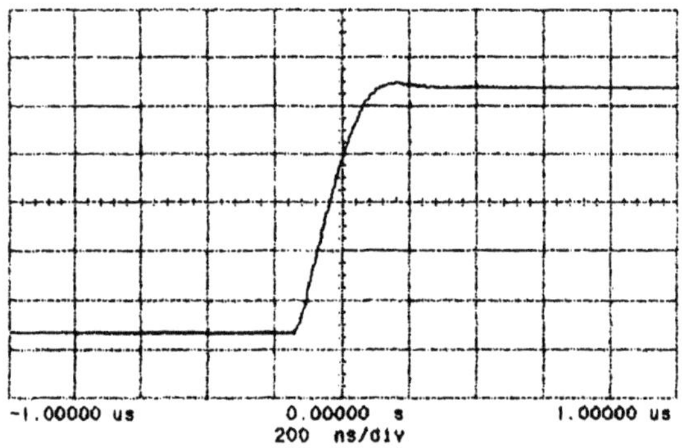

Fig. 5-35 *Response to a 1-V step of operational amplifier with rail-to-rail input stage.*

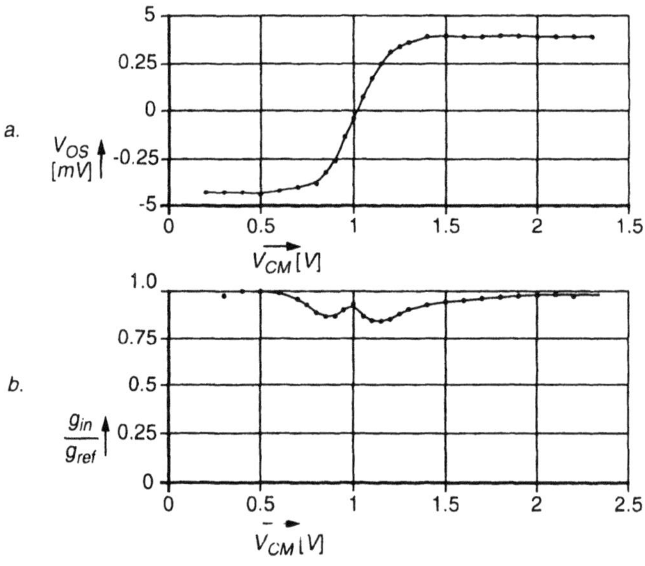

Fig. 5-36 *Offset voltage (a) and transconductance (b) as a function of common-mode input voltage of compact operational amplifier with rail-to-rail input stage.*

frequency vary approximately 18% which is slightly more than 13% which was calculated in chapter 2. An overview of the measurement results is given in Table5-4.

Table 5-4 *Measured specifications of compact operational amplifier with rail-to-rail input stage (V_{DD} = 1.8 V, T_A=25 °C, R_L=10 kΩ, C_L=5 pF).*

Parameter	Value	Unit
Die area	0.1	mm^2
Temp. range	-55 .. 150	°C
Min. supply voltage	1.7/2.5 [1])	V
Supply current	230	μA
Peak output current	1.2	mA
Input offset voltage	8	mV
Input noise voltage (f=100 kHz)	38	$nV/\sqrt{Hz}$
CM input range	V_{SS}-0.5 .. V_{DD}-1.3[1]) and V_{SS}+1.2 .. V_{DD}+0.5 [1])	V
CM output-voltage range	V_{SS}+0.2 .. V_{DD}-0.2	V
CMRR	40	dB
Unity-gain frequency	4	MHz
Phase margin	67	°
dc-gain	86	dB
Slew rate	4.5	V/μs
Settling time (1%, 1V)	370	ns
Settling time (0.1%, 1V)	465	ns

[1]) Rail-to-rail if V_{DD} > 2.5 V.

5.3.3 Ultimate-low-voltage compact amplifiers with rail-to-rail output ranges

The supply-voltage limitation of two gate-source voltages and one saturation voltage as found in the two-stage compact VLSI topology can only be surpassed by inserting at least one stage between the cascoding and summing circuit and the output stage. In the past, such an attempt has led to a four-stage ultimate low-voltage topology [20,21]. However, three stages are sufficient to realize an ultimate low-voltage circuit as demonstrated in principle in Fig. 5-37. The circuit is based on the two-stage amplifier principle depicted in 6-17. The circuit comprises the same output stage M_1, M_2 with

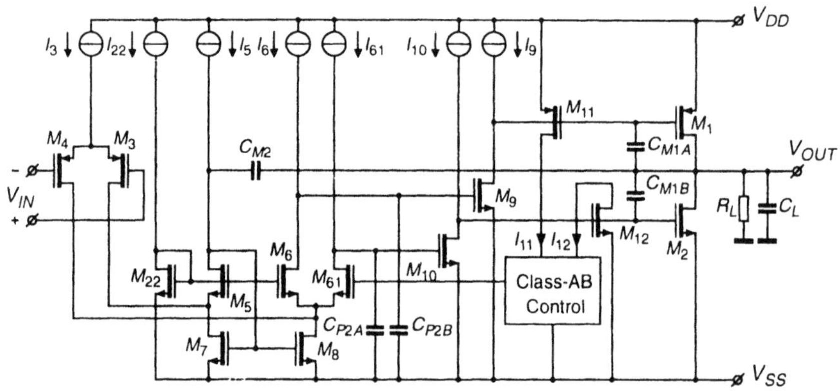

Fig. 5-37 *Principle of ultimate-low-voltage mirrored-nested-Miller topology.*

bias-current measuring transistors M_{11}, M_{12}, the same PMOS input stage M_3, M_4 with cascoding and summing circuit M_5-M_8, M_{16}. In between the summing circuit and the output stage, intermediate stage M_9, M_{10} has been inserted. Further, the upper half of the mesh of the circuit shown in fig6-17 consisting of M_9, M_{19} and I_6 have been replaced by two current source I_6, I_{16} because otherwise a stacking of two saturation voltages and one gate-source voltage would have occurred. For frequency compensation Miller capacitors C_{M1A} and C_{M1B} are placed across the output transistors. A second Miller capacitor required to compensate the three-stage circuit cannot simply be inserted across the output stage and the intermediate stage because both stages are inverting. Therefore mirrored-nested-Miller compensation is used consisting of outer Miller capacitors C_{M2A} and C_{M2B} and parallel capacitors C_{P2A} and C_{P2B}. Alternatively, a normal nested-Miller topology can be applied, but then current mirrors should be inserted between intermediate stage and output stage as shown in Fig. 5-38. The only supply-voltage limitation that occurs in both circuits shown in Fig. 5-37 and Fig. 5-38 is determined by a single gate-source voltage and a single saturation voltage. We need for example to bias the gate-source voltage of output transistor M_1 and a saturation voltage for proper operation of transistor M_9 in Fig. 5-37 or current source I_9 in Fig. 5-38. Also in other circuit parts we only need one gate-source voltage and one saturation voltage, for example in the summing circuit where a saturation voltage for proper operation of I_5 is needed on top of the gate-source voltage of M_7. This is in contrast with an earlier ultimate-low-voltage amplifier [20]. It is indeed true that the limitation first occurs in the output stage since the output stage should deliver considerable output currents. However, a this ultimate-low-voltage principle allows operation at a further reduced supply voltage, although the maximum output current is limited in the lowest part of the supply-voltage range. A final point of concern is the biasing of cascodes inside a single gate-

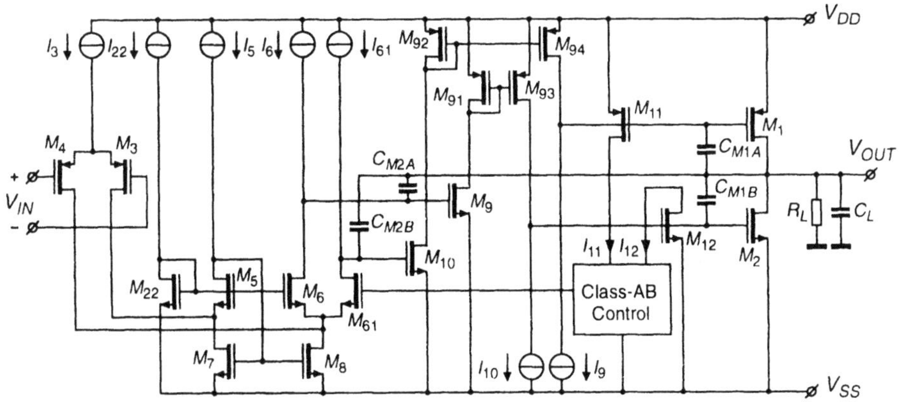

Fig. 5-38 *Principle of ultimate-low-voltage nested-Miller amplifier.*

source voltage. Currently, the threshold voltage of most processes is of the order of 0.7 V. This allows reasonable head room or biasing cascodes. However, the biasing of a cascode inside the gate-source voltage of an output-stage transistor, which is designed with large *W/L* ratio gives already problems over temperature [20]. If the threshold voltage further reduces it may be difficult to bias any cascode. We need, however, folded cascodes to obtain a reasonable input-voltage range. Is that also possible in the future? Of course, not all transistors are biased close to the threshold voltage. For example, the current-mirror transistors M_7, M_8 are usually biased well in strong inversion for reasons of noise and offset. Therefore, the gate-source voltage of M_7 and M_8 is much larger than the threshold voltage yielding sufficient head room for biasing cascode M_5 even if the threshold voltage is further reduced. Thus, we only have to take care that the intermediate stage transistors are biased in such a way that their current-source voltages are sufficient to allow transistors M_6 and M_{16} to function properly. This allows further reduction of the threshold voltage in future processes. Note that the biasing requirement for the intermediate stage agrees well with the basing condition for multipath nested Miller amplifiers where the transconductance of the intermediate stage should be small yielding a small *W/L* ratio. Also note that it is advantageous to apply NMOS transistors for transistors that operate in strong inversion since this yields the lowest gate-source voltage because the threshold voltage is usually smaller than the threshold voltage of PMOS transistors.

A complete implementation of the mirrored-nested-Miller ultimate-low-voltage amplifier is depicted in Fig. 5-39. The amplifier consists of output stage M_1, M_2, intermediate stage M_9, M_{10}, input stage M_3, M_4 and summing circuit M_5-M_8, M_{51}, M_{61}

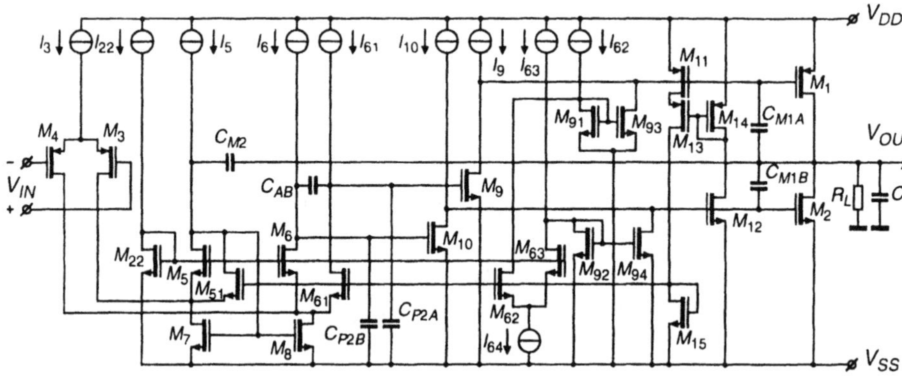

Fig. 5-39 *Complete implementation of mirrored-nested-Miller ultimate-low-voltage operational amplifier.*

according to the principle shown in Fig. 5-37. Class-AB control is created using a simple minimum selector structure consisting of M_{11}-M_{15}. The minimum selector drives class-AB control amplifier M_6, M_{16} which controls the input signal of the intermediate stage transistors. Thus, a feedback class-AB control is created. A second class-AB amplifier M_{62}, M_{63} implements a feedforward path and directly drives the output transistors M_1, M_2 via current mirrors M_{91}-M_{94} guaranteeing a stable class-AB control loop. The current mirrors are needed to maintain the low minimum supply voltage of one gate-source voltage and one saturation voltage. If the feedforward class-AB amplifier would directly be connected to the output stages the minimum supply voltage would be increased by a saturation voltage. By taking the W/L ratio of the current-mirror transistors small, sufficient headroom is available for biasing of the feedforward class-AB amplifier even using future processes with low threshold voltages. If necessary, the class-AB control loop can be further stabilized using capacitors C_{AB}.

As we see, the feedforward path of the class-AB control loop complicates the mirrored nested Miller circuit which is in principle elegantly simple. The class-AB feedforward path is, however, necessary since the control loop incorporates two stages, intermediate stage M_9, M_{10} and control amplifier M_6, M_{61} while hardly any means for compensation of the control loop can be employed. The best compensation means for the control loop is to insert Miller capacitors across M_9 and M_{10} but this would deteriorate the frequency behavior of the overall amplifier. The only compensation means left for controlling the class-AB loop is capacitor C_{AB}. But, without feedforward path the value of C_{AB} would be too large for integrated amplifiers. Also, the current mirrors required by the class-AB control loop cannot be used to create a compact feedforward path for the main amplifier because the summing circuit contains the signal of the outer Miller capacitor. The implementation of the feedforward path requires duplicating the input stage and the summing circuit as is demonstrated in the complete circuit diagram as shown in Fig. 5-40.

If nested Miller compensation is used instead of mirrored nested Miller compensation as shown in Fig. 5-41 a compact feedforward path can easily be created. The circuit is based on the principle depicted in Fig. 5-38. But, instead of using PMOS current mirrors between intermediate stage and output stage, NMOS current mirrors M_{91}-M_{94} are used which are also applied in the class-AB control loop. Frequency compensation is obtained using a nested-Miller structure comprising capacitors C_{M1A}, C_{M1B}, C_{M2A}, C_{M2B}. A compact multipath upgrades the compensation structure to a multipath-nested-Miller compensation. The multipath is created using transistors M_{71}, M_{81} which directly drive the output transistors M_1 and M_2, respectively. Transistors M_{71}, M_{81} are connected in parallel with the current mirror M_7, M_8 and copy out the input signal. The resulting multipath-nested-Miller amplifier has about the same complexity as the previous multipath-mirrored-nested-Miller amplifier.

The multipath-mirrored-nested-Miller (MMiNM) amplifier has been realized in a 1.6μm CMOS process with N-channel threshold voltage of 0.7V and P-channel threshold voltage of 1.1V. Therefore, the minimum supply voltage of this amplifier of 1.5 V is rather large. A micrograph of the opamp is shown in Fig. 5-42. The die area is 0.12 mm^2. The multipath-nested-Miller (MNMC) operational amplifier has been realized in 0.8μm BiCMOS with threshold voltages of 0.6V and 0.7V, respectively. The minimum supply voltage is as low as 1.2 V. A micrograph is shown in Fig. 5-42b and it occupies only 0.03 mm^2. Bode plots are shown in Fig. 5-43 and Fig. 5-44, respectively and specifications are listed in Table5-5. The MMiNM amplifier has a unity-gain frequency of 4.8 MHz and the MNM amplifier a unity gain frequency of 10 MHz. The response to a 0.5V step is plotted in Fig. 5-45 and Fig. 5-46, respectively.

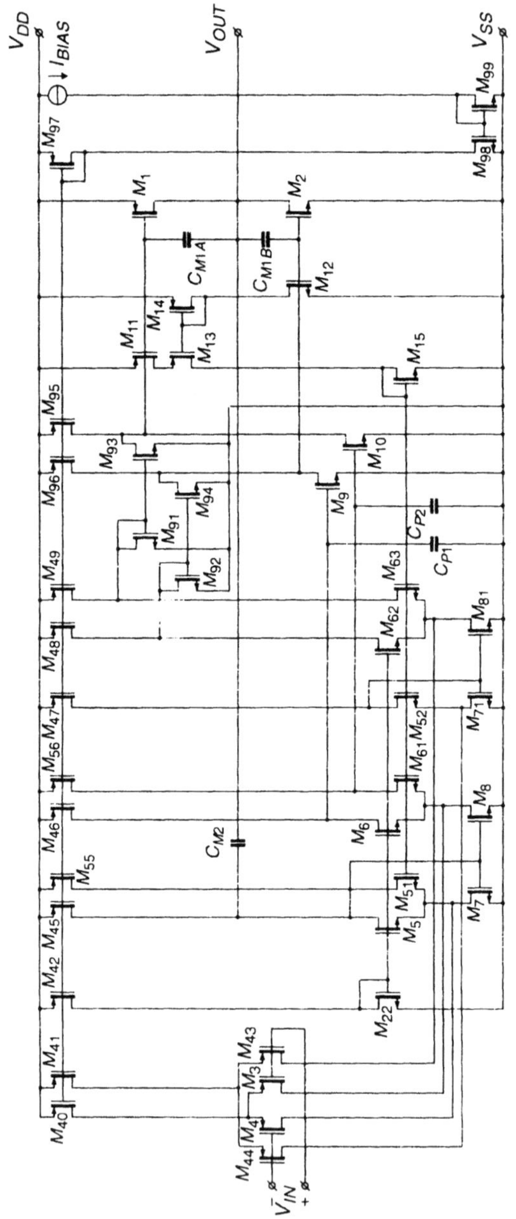

Fig. 5-40 *Complete circuit diagram of multipath mirrored nested Miller compensated ultimate-low-voltage operational amplifier.*

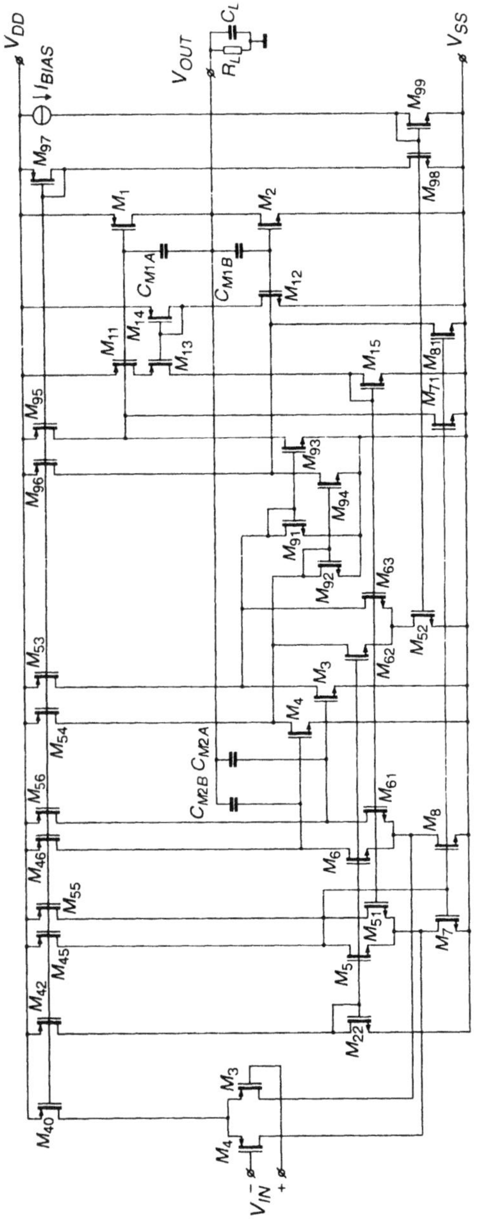

Fig. 5-41 *Circuit diagram of multipath-nested-Miller ultimate-low-voltage amplifier.*

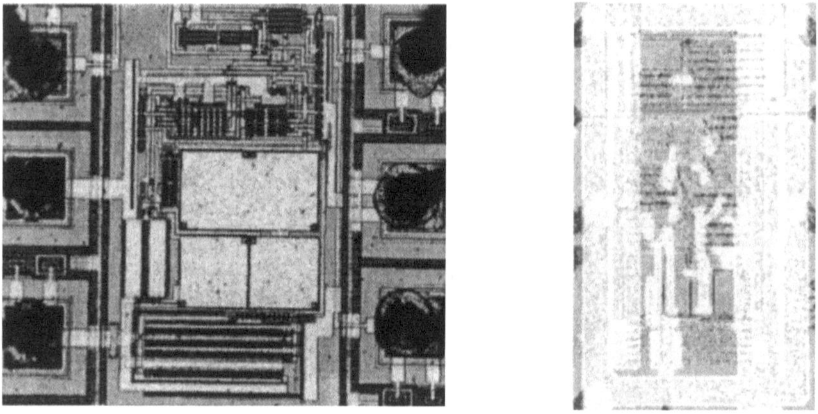

a. b.

Fig. 5-42 *Micrographs of ultimate-low-voltage operational amplifiers: multipath-mirrored-nested-Miller amplifier (a) and multipath-nested-Miller amplifier (b)*

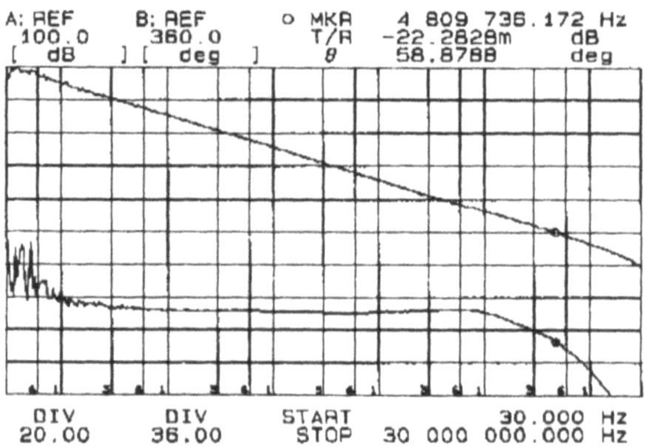

Fig. 5-43 *Frequency response of MMiNM amplifier*

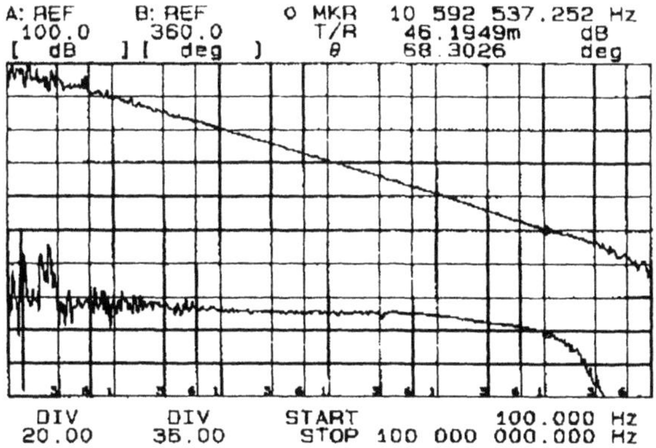

Fig. 5-44 *Frequency response of MNM amplifier.*

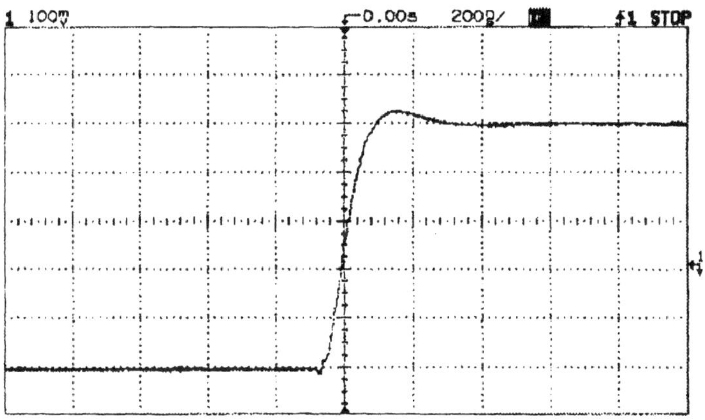

Fig. 5-45 *Response to 0.5V step of MMiNM amplifier.*

Table 5-5 *Measured specifications of compact operational amplifiers* (T_A=25 °C, R_L=10 kΩ, C_L=5 pF).

Parameter	MMiNM	MNM	Unit
Die area	0.12	0.03	mm^2
Temp. range	-55 .. 150	-55 .. 150	°C
Min. supply voltage	1.5	1.2	V
Supply current	206	437	μA
Peak output current	1.2	3	mA
Input offset voltage	5	5	mV
Input noise voltage (f=100 kHz)	40	50	$nV/\sqrt{Hz}$
CM input range	V_{SS}-0.6 .. V_{DD}-1.4	V_{SS}-0.5 .. V_{DD}-1.2	V
CM output-voltage range	V_{SS}+0.2 .. V_{DD}-0.2	V_{SS}+0.2 .. V_{DD}-0.2	V
CMRR	80	80	dB
Unity-gain frequency	4.8	10	MHz
Phase margin	59	68	°
dc-gain	97	98	dB
Slew rate	4	8	$V/\mu s$
Settling time (1%, 1 V)	420	240	ns
Settling time (0.1%, 1 V)	610	400	ns

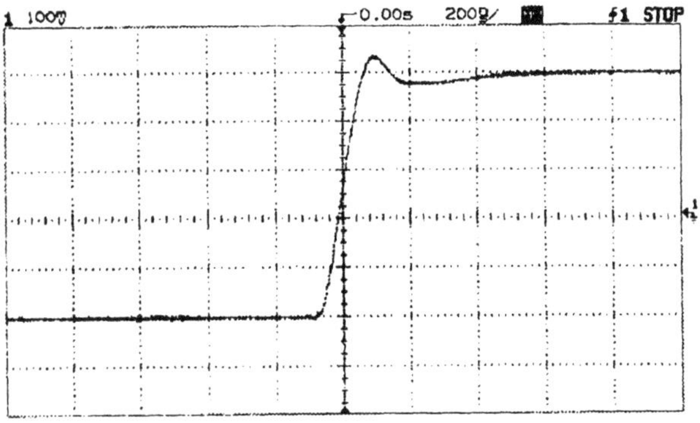

Fig. 5-46 *Response to 0.5V step of MNM amplifier.*

5.4 BiCMOS amplifier topologies

In the previous section amplifier topologies have been presented using bipolar technology or CMOS technology. But, what can we do if BiCMOS technology is available? Is it possible to take advantage of the high bipolar transconductance and the high CMOS input impedance, and combine these properties into a simple circuit? In this section we show that it is possible to combine bipolar and CMOS transistors yielding a combination of high gain and high bandwidth.

5.4.1 Compact BiCMOS amplifiers with rail-to-rail input and output ranges

The first possibility to use BiCMOS is based on the compact CMOS amplifier topology discussed in section 5.3.1. A BiCMOS version is depicted in Fig. 5-47. In this circuit the CMOS input-stage transistors are replaced by bipolar transistors Q_3, Q_4 in order to obtain low offset and noise voltage. The input stage drives the output stage M_1, M_2 via the same cascoding and summing circuit M_5-M_9, M_{15}, M_{16}, M_{19} as discussed in section 5.3.1. Also, the class-AB control circuit M_{10}-M_{14}, M_6, M_{16} is identical. Using the high transconductance of bipolar transistors the bandwidth of the amplifier could be improved by replacing the output-stage transistors by bipolar transistors. But, due to the low input impedance of bipolar transistors such a design would have low dc gain. An alternative is to place a bipolar output stage in parallel with the CMOS class-AB output stage as depicted in Fig. 5-47. The bipolar output transistor Q_{21} is compensated using Miller capacitor C_{M1C} and is driven by an independent feedforward path created by

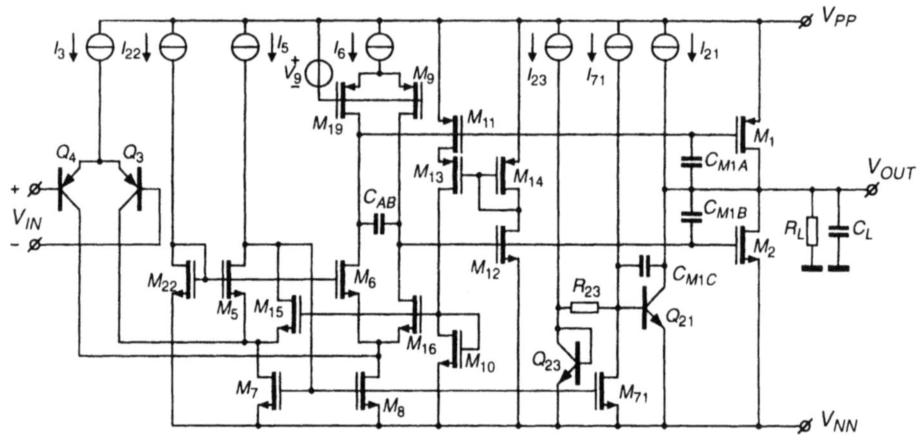

Fig. 5-47 *Compact two-stage BiCMOS amplifier with multipath-parallel-Miller compensation.*

transistor M_{71}. In fact this is another form of a multipath technique which is not yet discussed and is called multipath-parallel-Miller compensation. A flat characteristic without pole-zero doublets is obtained if the unity-gain frequency of the bipolar parallel path is equal to the bandwidth of the main amplifier. This is obtained if all Miller capacitors are equal.

Then, the bandwidth limiting second pole is given by

$$\omega_2 = \frac{g_1 + g_2 + g_{21}}{C_L} \tag{5-17}$$

which clearly shows the improvement of the bandwidth by the transconductance g_{21} of the bipolar output transistor. Somehow, we need to force a bias current through Q_{21}. The bias current of Q_{21} is forced by transistor Q_{23} and current source I_{23}. Resistor R_{23} is inserted to limit the input impedance reduction of the bipolar output transistor by Q_{23} so that the unity-gain frequency of the bipolar amplifier is still accurately determined by C_{M1C}. Thus, the bipolar parallel stage can be used to increase the capacitive loading capability of the CMOS output stage or to reduce the bias current for a certain bandwidth. Using the pure two-stage technology it is not possible to use bipolar transistors to increase the output current capability. The amplifier can operate on supply voltages of one gate-source voltage and two saturation voltages which amounts to 1.5-2.0 V depending on technology.

At higher supply voltages a BiCMOS Darlington transistor can be used to immensely increase the output current capability. An amplifier employing the BiCMOS Darlington as shown in Fig. 5-48 is a good example of combining bipolar and CMOS design strategies. The amplifier consists of a bipolar rail-to-rail input stage Q_5, Q_6, Q_{51}, Q_{61}, a

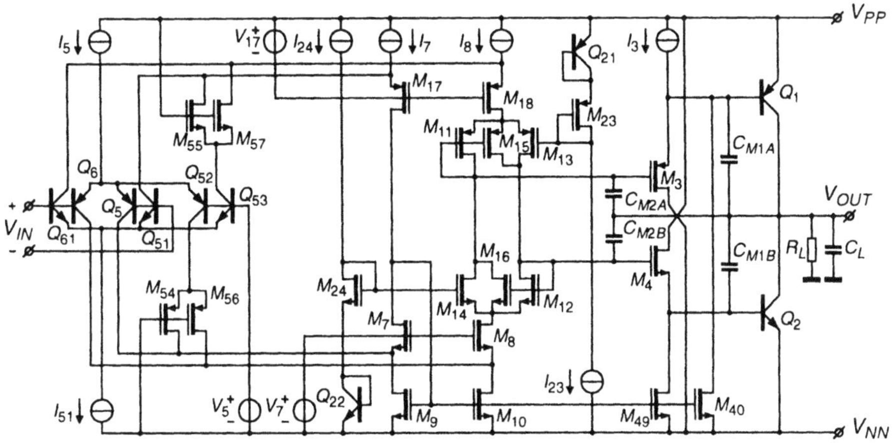

Fig. 5-48 *Compact quasi-two-stage amplifier with BiCMOS Darlington output stage, multipath-nested-Miller compensation and rail-to-rail input stage.*

CMOS cascoding and summing circuit M_7-M_{10}, M_{17}, M_{18} and BiCMOS Darlington output stage Q_1, M_3, Q_2, M_4. Using CMOS technology to implement the cascodes M_7, M_8, M_{17}, M_{18} and Darlington transistors M_3 and M_4 a very high impedance is created at the input of the output stage yielding high dc gain. By driving the bipolar output transistor by a CMOS driver, a large output current can be supplied while the bias currents of the input stage and the summing circuit can still be chosen small yielding high power efficiency. By employing a bipolar rail-to-rail input stage, a rail-to-rail input voltage range is achieved while the offset and noise voltage are small. Another important advantage of the bipolar rail-to-rail input stage over a CMOS implementation is that the common-mode rejection is much better, of the order of 60 *dB* because of the low offset of bipolar transistors. The transconductance of the rail-to-rail input stages is regulated constant using switches Q_{52}, Q_{53}. The current flowing through the switches is added to the currents of the input-stage transistors by M_{54}-M_{57} so that the output currents of the input stage are independent of the common-mode input voltage. Therefore, the cascoding and summing circuit can be realized using a simple structure comprising cascodes M_7, M_8, M_{17}, M_{18} current sources I_7, I_8 and, current mirror M_9, M_{10}. Because the current switches are in parallel with the input stages, the input noise of the amplifier is increased. By supplying the current of the current switches via CMOS transistors M_{54}-M_{57} the noise contribution of the g_m-control circuit is small because the transconductance of CMOS transistors M_{54}-M_{57} is much smaller than the transconductance of input stages Q_5, Q_6 and Q_{51}, Q_{61}. Due to the relatively large supply voltage required by the Darlington transistors there is sufficient headroom so that the class-AB biasing of the output stage can accurately be controlled using a feedforward control consisting of M_{11}-M_{16}, Q_{21}, Q_{22}, M_{23}, M_{24}. The feedforward control does not present any stability problems such as with feedback control. The biasing of the output transistors is strongly controlled using two translinear loops consisting of Q_1, M_3, M_{11}, M_{13}, M_{23}, Q_{21} and Q_2, M_4, M_{12}, M_{14}, M_{24}, Q_{22}. Heart of the control circuit is mesh M_{11}-M_{14} to which transistors M_{15} and M_{16} have been added to assure a minimum current in the output transistors. The control circuit is a BiCMOS version of the bipolar circuit discussed in chapter 3, section 3.4. Although the circuit has a two-stage topology with respect to the complexity of the design, the frequency compensation is a different story. Because of the small transconductance of MOS transistors the intermediate poles of the Darlington transistors at the sources of M_3 and M_4 are more prominent than the intermediate poles in an entire bipolar Darlington. Therefore, with regard to frequency compensation the amplifier should be considered as a three-stage design. The three-stage topology requires nested-Miller compensation comprising capacitors C_{M1A}, C_{M2A} and C_{M1B}, C_{M2B}. Fortunately, the multipath technique can be easily applied. The input signal of both the NPN and the PNP input stage is present in current mirror M_9, M_{10}. Therefore we can create a copy of the input signal by placing additional transistors M_{40}, M_{49} with their gate and source terminals in parallel with the gate and source terminals of M_9 and M_{10}. Thus, the multipath is implemented without duplicating the complete input stage and summing circuit while using only two transistors M_{40}, M_{49}! Using the multipath-nested-Miller compensation the CMOS transistors M_3 and M_4 are bypassed for high frequencies and the bandwidth of the bipolar output transistors is fully exploited. The amplifier can operate at a supply voltage of about 3 V depending on technology, has a compact structure and can deliver large currents of the order of 100 mA if the supply voltage is not too small while the standby current can be as low as 100 μA.

If we want to take advantage of BiCMOS at lower supply voltages, the BiCMOS Darlington cannot be used. A first attempt to create a compact BiCMOS amplifier at lower supply voltages is depicted in Fig. 5-49. The compact three-stage BiCMOS operational amplifier has basically the same structure as the previous design containing a bipolar rail-to-rail input stage Q_5, Q_6, Q_{51}, Q_{61}, a CMOS cascoding and summing circuit M_7-M_{10}, M_{17}, M_{18} and a low voltage version of the BiCMOS Darlington output stage Q_1, M_3, Q_2, M_4. Transistors M_3 and M_4 now form a plain intermediate stage that drives output transistors Q_1, Q_2 via current mirrors Q_{31}, Q_{32} and M_{41}, M_{42}. The current mirrors are implemented using NPN and PMOS transistors because they display better behavior than NMOS transistors and PNP transistors, respectively, when driving large base

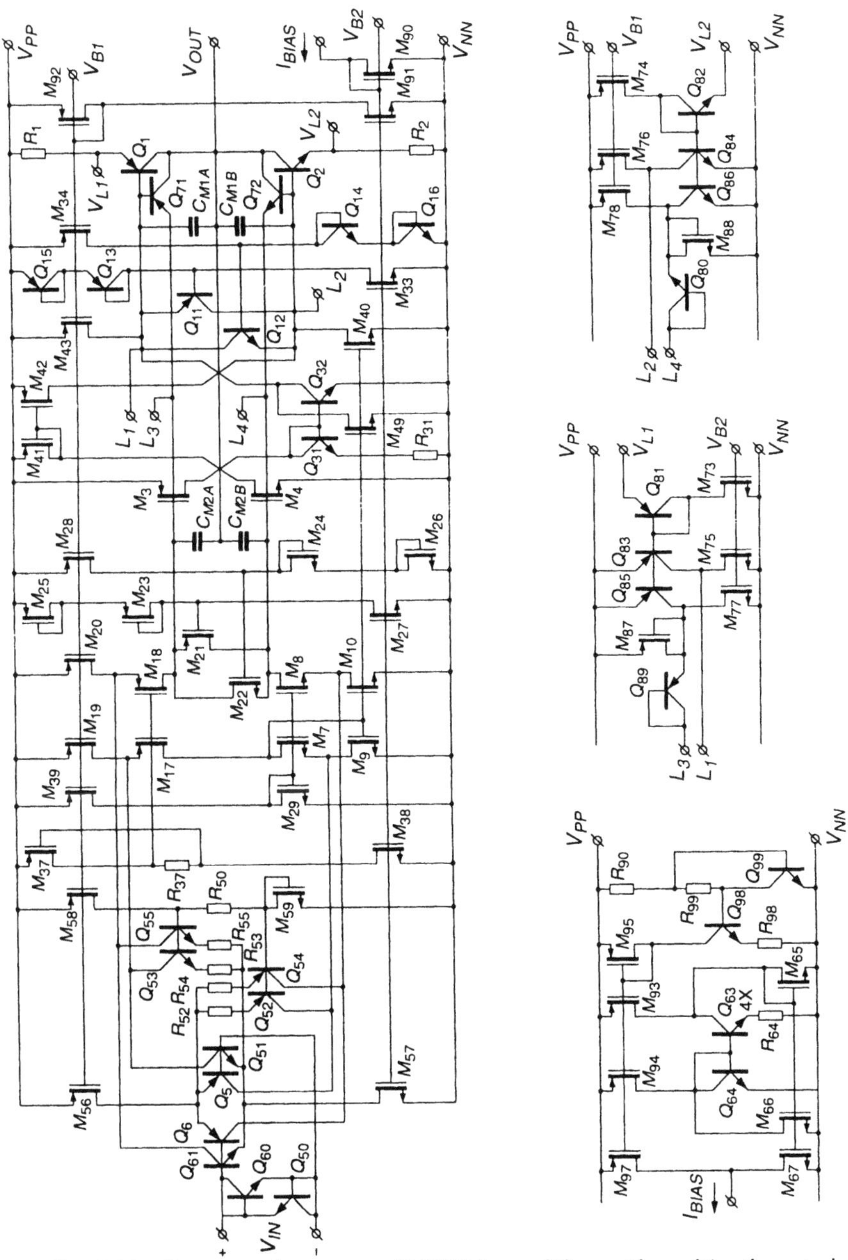

Fig. 5-49 *Compact three-stage BiCMOS amplifier with multipath-nested-Miller compensation, current limiters and PTAT source.*

currents. Class-AB biasing of the output transistors Q_1, Q_2 is created by a feedforward control consisting of Q_{11}-Q_{16}. The biasing of the output transistors is strongly controlled by two translinear loops Q_1, Q_{11}, Q_{13}, Q_{15} and Q_2, Q_{12}, Q_{14}, Q_{16}. The mesh transistors Q_{11}, Q_{12} guarantee a minimum current in the output transistors as long as we do not supply the base currents of the output stage via the mesh transistors. Therefore, the collector of Q_{32} is directly connected to the base of Q_1 and the drain of M_{42} is connected to the base of Q_2. Thanks to the feedforward class-AB control we need not worry about the stability of a class-AB feedback loop and the structure of the circuit remains straightforward. But there are two side effects. First of all, the minimum supply voltage is not much smaller than the minimum supply voltage of the previous amplifier. Secondly, if we drive both output transistors by an intermediate-stage transistor, which is preferable in order to supply sufficient base current, we have to fix the bias current of the intermediate stage. In this example the bias current of the intermediate stage is determined using a similar circuit as for the class-AB biasing of the output stage consisting of transistors M_{21}-M_{26}. This way, two translinear loops M_3, M_{21}, M_{23}, M_{25} and M_4, M_{22}, M_{24}, M_{26} control the biasing of the intermediate stage M_3, M_4. The transconductance of the intermediate stage is regulated by a similar current-switch-based g_m-control circuit as in the previous design comprising switches Q_{52}-Q_{55}. However, in this example resistors R_{52}-R_{55} are used to reduce the transconductance of the switches and limit their contribution to the input referred offset and noise voltage. Capacitors C_{M1A}, C_{M2A} and C_{M1B}, C_{M2B} take care of the frequency compensation while transistors M_{40}, M_{49} create a feedforward path. Using this multipath-nested-Miller topology, again, the high gain of the three stage BiCMOS amplifier is combined with the high bandwidth of a two-stage bipolar amplifier. Another version of this amplifier is depicted in fFig. 5-50. Instead of the normal nested Miller compensation scheme as used in the previous design, the outer Miller capacitors are now connected via cascodes to the intermediate stage yielding a multipath-cascoded-nested-Miller compensated amplifier. Both BiCMOS amplifiers are currently being realized in a $0.8\mu m$ BiCMOS process. The simulated specifications of these amplifiers are listed in Table5-6. The amplifiers can operate down to a supply voltage of two gate-source voltages and one saturation voltage which is about $2.5V$, depending on technology. The minimum supply voltage is therefore not much smaller than the minimum supply voltage of the previously discussed quasi-two-stage amplifier. But, using the same transistor sizes and the same supply voltage it can deliver a larger output current due to the low-voltage Darlington structure.

The low-voltage Darlington structure can readily function at lower supply voltages if we do not apply feedforward class-AB biasing. This is discussed in the next section.

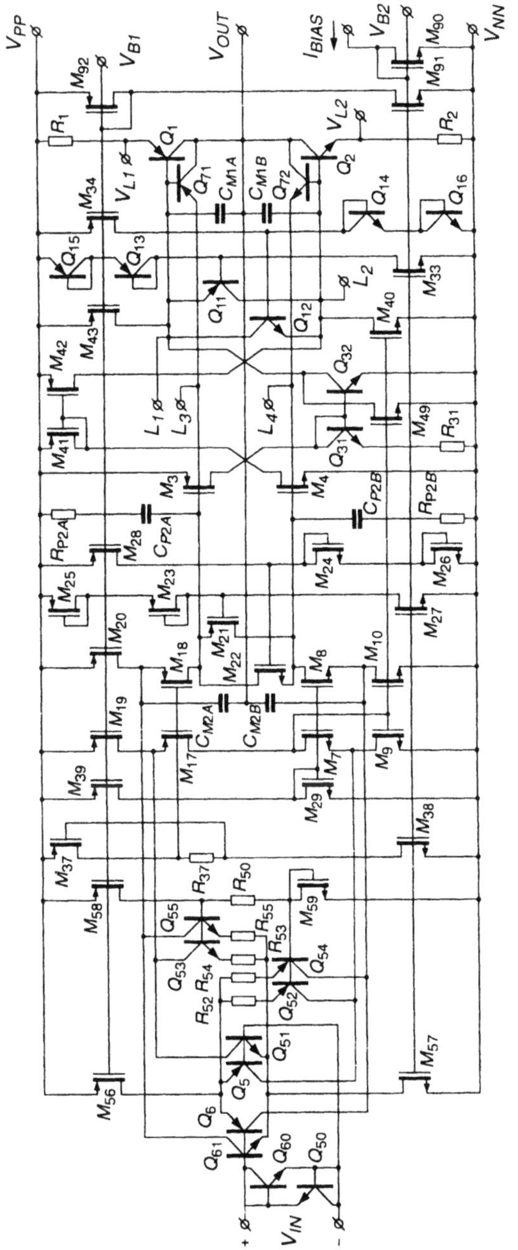

Fig. 5-50 *Compact three-stage BiCMOS operational amplifier with multipath-cascoded-nested-Miller compensation.*

Table 5-6 *Specifications of compact BiCMOS operational amplifiers (T_A=25 °C, R_L=10 kΩ, C_L=5 pF).*

Parameter	MNM	MCNM	Unit
Die area	0.1	0.1	mm^2
Temp. range	-55 .. 150	-55 .. 150	°C
Min. supply voltage	2.5	2.2	V
Supply current	0.95	0.95	mA
Peak output current	+6.5/-9	+6.5/-9	mA
Input offset voltage	1	1	mV
Input noise voltage (f=100 kHz)	15	15	nV/√Hz
CM input range	V_{NN}-0.3 .. V_{PP}+0.3	V_{NN}-0.3 .. V_{PP}+0.3	V
CM output-voltage range	V_{SS}+0.1 .. V_{DD}-0.1	V_{SS}+0.1 .. V_{DD}-0.1	V
CMRR	90	90	dB
Unity-gain frequency	20	20	MHz
Phase margin	62	62	°
dc-gain	105	105	dB
Slew rate	6	6	V/µs

5.4.2 Low-voltage BiCMOS amplifiers with rail-to-rail output range

The low-voltage BiCMOS Darlington circuit, which is in fact a cascade of an MOS inverting amplifier stage and a bipolar inverting amplifier stage interconnected by a current mirror, allows operation at supply voltages well below 2*V*. To reach such a low supply voltage, feedforward class-AB control cannot be applied to accurately control the output stage and feedback biasing should be used. The resulting low-voltage three-stage BiCMOS operational amplifier is presented in Fig. 5-51. The amplifier consists of bipolar input stage Q_5, Q_6, summing circuit M_7-M_{10}, M_{17}-M_{19}, M_{29}, intermediate stage M_3, M_4 and output stage Q_1, Q_2. The combination of input stage, summing circuit and intermediate stage is identical to the first amplifier shown in Fig. 5-47 without the bipolar parallel output stage and the class-AB control. The intermediate stage M_3, M_4 drives the output transistors via current mirrors Q_{31}, Q_{32} and M_{41}, M_{42}. Class-AB biasing of Q_1 and Q_2 is implemented using a feedback loop consisting of measuring circuit and

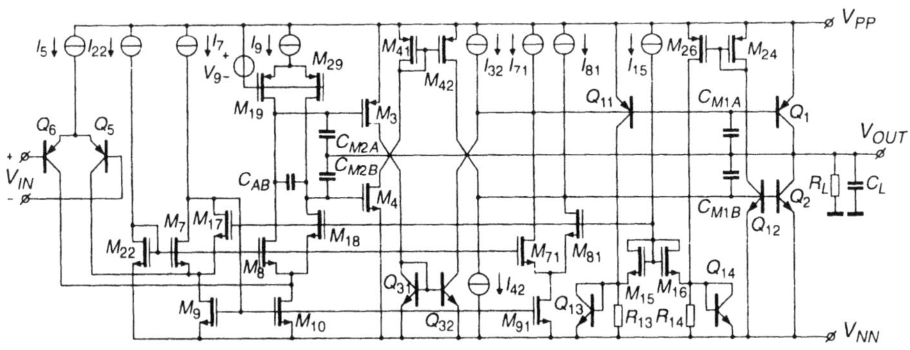

Fig. 5-51 *Low-voltage three-stage BiCMOS operational amplifier with rail-to-rail.*

minimum selector Q_{11}, Q_{12}, M_{15}, M_{16}, M_{24}, M_{26} and class-AB amplifier M_8, M_{18} which is part of the mesh M_8, M_{18}, M_9, M_{19} which drives the intermediate stage. The bias current of Q_1 is measured by Q_{11} while the bias current of Q_2 is measured using Q_{12}. The collector current of Q_{11} generates a voltage drop across R_{13} that is proportional to the collector current of Q_1. The collector current of Q_{12} flows through current mirror M_{24}, M_{26} and generates a voltage across R_{14} that is proportional to the collector current of Q_2. The output of minimum selector M_{15}, M_{16} is controlled by the lower of the two voltages across R_{13} and R_{14} which corresponds to the lower of the bias currents of the output transistors. Diode-connected transistors Q_{13}, Q_{14} limit the voltage across R_{13} and R_{14}. The output voltage of the minimum selector drives the class-AB amplifier M_8, M_{18} which completes the feedback loop by driving the intermediate stage. Thus, class-AB biasing is obtained with a guaranteed minimum current in the output stage. The minimum selector can only be implemented using NMOS transistors because of the matching with NMOS transistors M_7, M_8, M_{17}, M_{18}, M_{71}, M_{81}. The matching is important to accurately control the drain-source voltages of M_9, M_{10}, M_{91} for assuring correct operation of these transistors. The stability of the feedback class-AB control loop is tackled by inserting a feedforward path consisting of differential pair M_{71}, M_{81} that is driven by the minimum selector and directly steers the output transistors Q_1, Q_2. The overall amplifier is compensated by a multipath-nested-Miller compensation structure consisting of capacitors C_{M1A}, C_{M2A} and C_{M1B}, C_{M2B} and a multipath created by M_{91} that drives the output transistors via transistors M_{71}, M_{81}. Since the compensation of the overall amplifier is not designed to compensate the class-AB feedback loop, the stability problem of the class-AB loop is difficult to solve using only a feedforward path. To improve the stability of the class-AB control loop, capacitor C_{AB} can be inserted. The amplifier can operate down to a supply voltage of one gate-source voltage and two saturation voltages which amounts to 1.5 - 2 V depending on technology. If necessary the supply voltage can be further reduced by applying techniques as described in section 5.3.3. Cascodes M_{19} and M_{29} should be removed and current mirrors should be inserted between the feedforward class-AB amplifier M_{71}, M_{81} and the output stage. However, the dc gain reduces by removing cascodes M_{19}, M_{29}.

Based on the low-voltage BiCMOS topology two operational amplifiers have been implemented in 0.8μ*m* BiCMOS. The photomicrographs of the realized amplfiiers are shown in Fig. 5-52. The circuit diagram of the first amplifier is depicted in Fig. 5-53. It

a.

b.

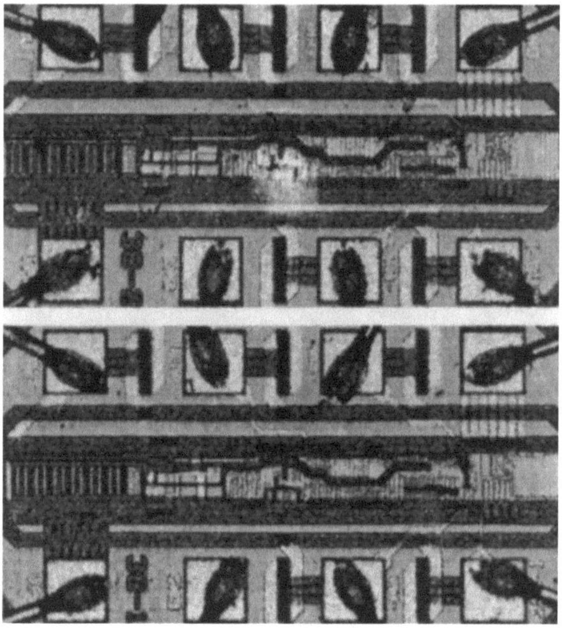

Fig. 5-52 *Photomicrographs of BiCMOS low-voltage operational amplifiers: (a) NM amplfier and (b) MNM amplifier.*

does not contain a feedforward path yielding a standard nested-Miller (NM) topology. The second amplifier is depicted in Fig. 5-54 and employs the multipath-nested-Miller (MNM) topology. The frequency response of the amplifiers is shown in Fig. 5-55. The standard version achieves a unity-gain frequency of 18*MHz* while the multipath version has a unity-gain frequency of 26*MHz*. The response to a 1*V* step is depicted in Fig. 5-56. The NM amplifier settles to 1% within 100 *ns* and to 0.1% within 260 *ns*. The MNM

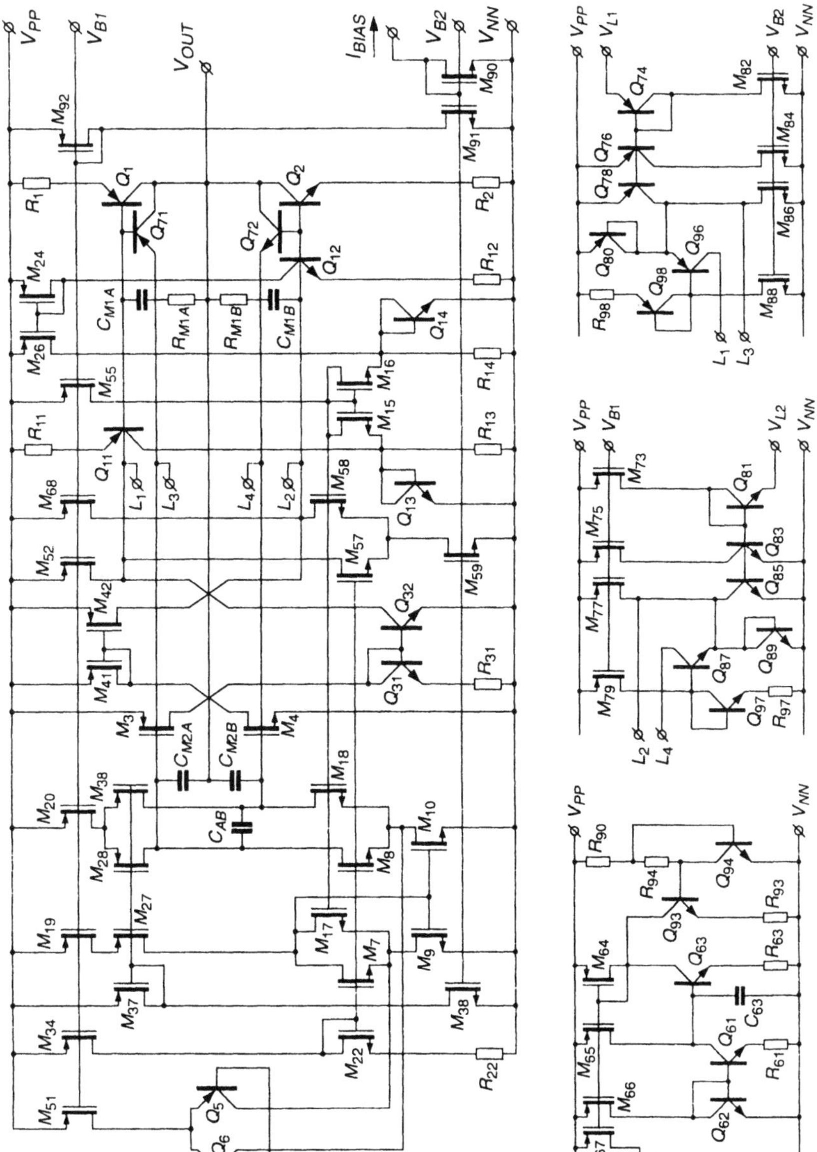

Fig. 5-53 *Low-voltage NM BiCMOS operational amplifier with rail-to-rail output range.*

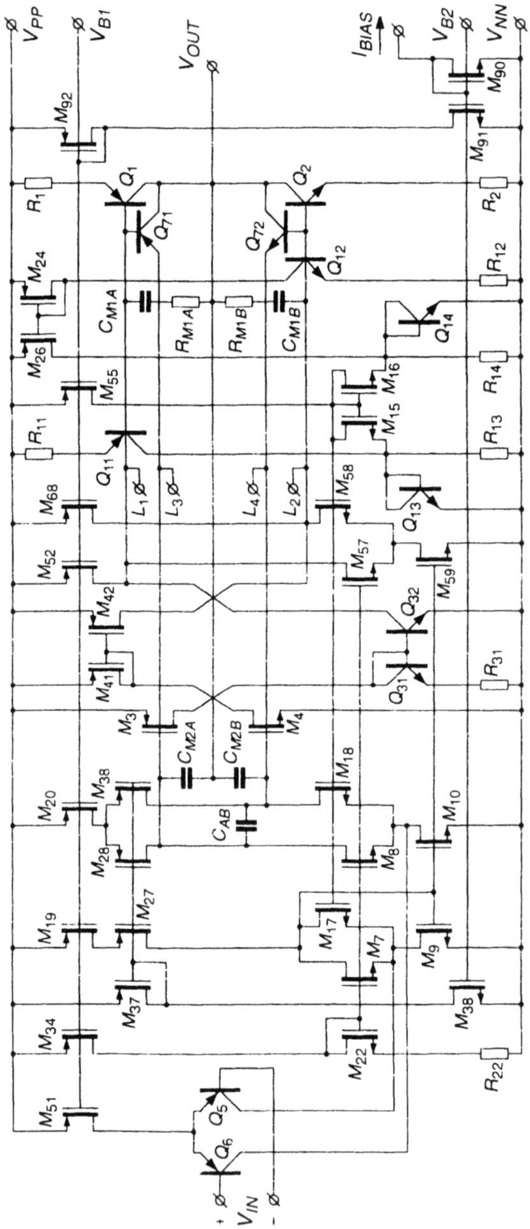

Fig. 5-54 *Complete circuit diagram of low-voltage MNM BiCMOS operational amplifier with rail-to-rail output range.*

a.

b.

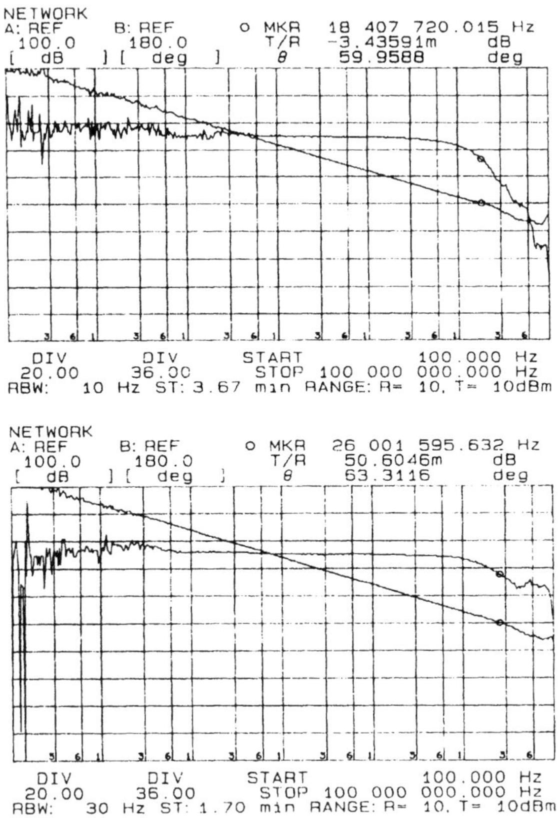

Fig. 5-55 *Frequency response of low-voltage three-stage BiCMOS operational amplifiers: (a) NM amplfier and (b) MNM amplifier.*

amplifiers settles to 1% within 100 *ns* and to 0.1% within 300 *ns*. Both amplifiers can operate on a supply voltage down to 1.3 *V*. An overview of the measured specifications is given in Table5-7.

Table 5-7 *Measured specifications of low-voltage BiCMOS operational amplifiers (T_A=25 °C, R_L=10 kΩ, C_L=5 pF).*

Parameter	NM	MNM	Unit
Die area	0.1	0.1	mm^2
Temp. range	-55 .. 150	-55 .. 150	°C
Min. supply voltage	1.3	1.3	V
Supply current	1.2	1.2	mA
Peak output current	+3.5/-7	+3.5/-7	mA
Input offset voltage	1	1	mV
Input noise voltage (f=100 kHz)	12	12	$nV/\sqrt{Hz}$
CM input range	V_{NN}-0.3 .. V_{PP}-1	V_{NN}-0.3 .. V_{PP}-1	V
CM output-voltage range	V_{SS}+0.1 .. V_{DD}-0.1	V_{SS}+0.1 .. V_{DD}-0.1	V
CMRR	90	90	dB
Unity-gain frequency	18	26	MHz
Phase margin	70	70	°
dc-gain	110	110	dB
Slew rate	7	8	V/μs
Settling time (1%, 1V)	100	100	ns
Settling time (0.1%, 1V)	260	300	ns

5.4.3 200-MHz power-efficient BiCMOS amplifiers

In most BiCMOS processes only lateral PNP transistors are available which have a performance that is by far inferior to NPN transistors. Even if vertical PNP devices are available their behavior is considerably worse. The size of a PNP is much larger for delivering the same current while the current gain is significantly smaller. Therefore PNP transistors cause major problems if applied in output stages that should deliver large output currents. Further, the PNP transit frequency f_T is at least a factor three smaller than the transit frequency of NPN transistors. Again, this mainly gives rise to limitations if the PNP transistors are applied in the output stage. In the quiescent state the output-stage transistors are biased at a relatively small current so that the actual transit frequency is much smaller than the peak value. If the previous bipolar rail-to-rail stages are for example implemented in a technology with a PNP peak f_T of 200 *MHz* the

a.

b.

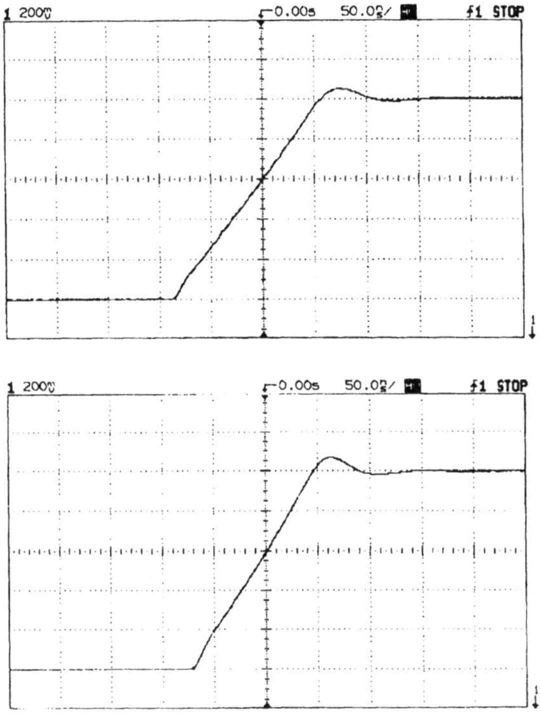

Fig. 5-56 *Response to a 1V step of BiCMOS low-voltage operational amplifiers: (a) NM amplfier and (b) MNM amplifier.*

bandwidth of the amplifier is limited to 15 to 20 *MHz*! For higher bandwidths or for driving high output currents, the all NPN output stage should therefore be applied. In combination with PMOS drivers a very powerful circuit is created that can be used for high power efficiency or for combining very high bandwidths with high dc gain. Of course the drawback is a slightly reduced output voltage range which is about one base-emitter voltage smaller than the output voltage range of a rail-to-rail output stage. But a bandwidth improvement of a factor of ten can easily make up for the reduced output-voltage range! A three-stage implementation of a BiCMOS operational amplifier with all-NPN output stage is depicted in Fig. 5-57. Although the circuit is somewhat complicated at a first glance, the basic topology is straightforward. The amplifier consists of a PNP input stage Q_5, Q_6 that drives PMOS intermediate-stage transistors M_3, M_4. The input stages consists of PNP transistors so that the common-mode input voltage range is similar to the common-mode output voltage range of the all-NPN output stage. The intermediate stage directly drives all-NPN output stage Q_1, Q_2. The remaining part

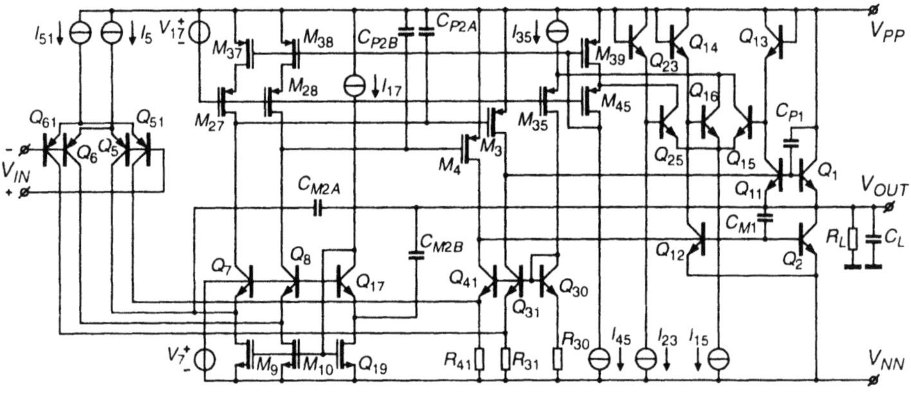

Fig. 5-57 *High-frequency power-efficient three-stage BiCMOS operational amplifier using all-NPN output stage.*

of the circuit is required to implement class-AB biasing of the output stage and to frequency compensate the amplifier. Class-AB biasing is obtained using a feedback control loop comprising transistors Q_{11}-Q_{16}, Q_{25}, M_{37}-M_{39}, M_{27}, M_{28}, M_{45}. The base-emitter junctions of Q_{11}, Q_{12} are connected in parallel with the output transistors Q_1, Q_2 so that the collector currents of Q_{11}, Q_{12} are proportional to the bias currents of the output stage. The collector currents of Q_{11} and Q_{12} generate voltage drops across Q_{13} and Q_{14} that drive minimum selector Q_{15}, Q_{16}. The output of the minimum selector is driven by the smaller of the voltage drops which corresponds to the smaller of the output-stage bias currents. Together with Q_{25} the minimum selector forms a class-AB amplifier of which the two outputs, the collector of Q_{25} and the sum of the collectors of Q_{15} and Q_{16}, are both used. The first output, the collector of Q_{25}, drives the intermediate stage via current mirror M_{37}-M_{39} and cascodes M_{27}, M_{28} completing the class-AB control loop. Thus, the minimum current of the output stage is controlled. The second output of the class-AB amplifier directly drives the bipolar output stage via cascode M_{35} and current mirror Q_{30}, Q_{31}, Q_{41} creating a feedforward path that guarantees stability of the class-AB control loop. Frequency compensation of the output stage is obtained using C_{M1} and C_{P1}. The three-stage circuit requires a nested-Miller structure, but due to the inverting intermediate stage a special configuration is needed. The upper half of the amplifier M_3, Q_1 does not pose any problems since M_3 is an inverting stage while Q_1 is a non-inverting stage. Therefore, a nested Miller capacitor can be inserted from the output node to the gate of M_3. In this case the Miller capacitor C_{M2A} is connected cascode Q_7 which drives the gate of M_3. Together with C_{P2A} a cascoded-nested-Miller structure is used to compensate the upper half of the amplifier. The lower half of the amplifier consists of two inverting stages M_4, Q_2 and cannot be compensated using conventional-nested-Miller compensation. It requires an additional inversion created by current mirror M_{19}, M_{10}. Thus, a mirrored-nested-Miller topology is used to compensate the lower half of the circuit consisting of C_{M2B}, M_{19}, M_{10} and C_{P2B}. Capacitors C_{P2A} and C_{P2B} assure that the high-frequency gain of the cascode and the current mirror in both nested Miller loops is of the order of unity avoiding complex poles. For reasons of noise the bias current of cascode Q_7 should be generated by the same circuit as the bias current of Q_8 using M_9. Drawback is that the signal of capacitor C_{P2B} now also flows to the upper half of the amplifier. This, however, can be compensated by making C_{P2A} two times larger than C_{P2B}. Capacitor C_{M2B} is connected to current mirror M_{19}, M_{10} via cascode Q_{17} so that the capacitor does not load the current mirror. A feedforward path is created using additional input stage Q_{51}, Q_{61} that directly drives the output stage via folded cascodes Q_{31}, Q_{41}. This way the PMOS intermediate stage is bypassed for high frequencies so that the bandwidth of the bipolar output stage is fully exploited. The amplifier can operate at supply voltages down to 2 *V* while achieving high bandwidths of more than 100 *MHz*. Alternatively, the design is very suited for driving large output currents of the order of 100 *mA* at lower bandwidths.

The extraordinary topology of the all-NPN output stage comprising a follower and an inverter enables different compensation schemes as shown in Fig. 5-58. The amplifier is

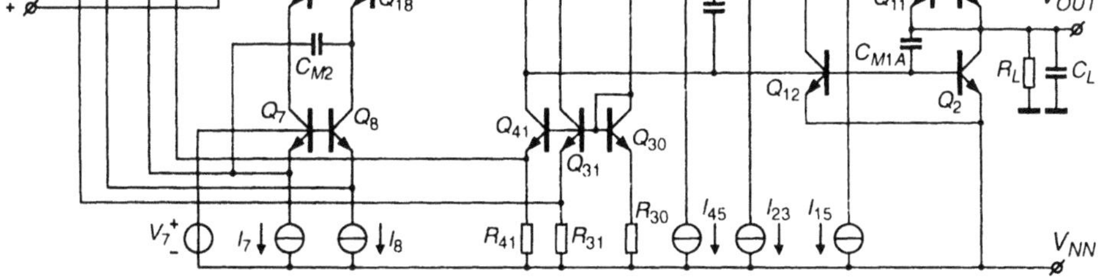

Fig. 5-58 *High-frequency power-efficient three-stage BiCMOS operational amplifier demonstrating alternative compensation scheme when using all-NPN output stage.*

nearly identical to the previous circuit. The Miller capacitors are connected in a different way, the current mirror of the mirror-nested-Miller structure is removed and cascodes Q_{17}, Q_{18} have been added to accommodate the new compensation scheme. A small Miller capacitor C_{M1A} is used to balance the output stage and should be equal to the parasitic capacitance connected to the base of Q_1. But the main compensation of the output stage is performed by C_{M1B} which is connected between the bases of the two output transistors. Because Q_1 is operating as a follower the output voltage is also present across C_{M1B}. Therefore the signal supplied by C_{M1B} to the bases of Q_1 and Q_2 is equal to the signal supplied by C_{M1} and C_{P1} in the previous amplifier. Thus, the function of C_{M1} and C_{P1} in the previous amplifier is the same as C_{M1B}. By replacing the two capacitors by a single capacitor about a factor of two in capacitive area is saved. However, the second pole at the output after insertion of C_{M1B} is given by

$$\omega_2 = \frac{g_{m2}}{C_L} \qquad (5\text{-}18)$$

which is a factor of two smaller than the second pole of normal Miller compensation. Therefore, frequency compensation using floating capacitor C_{M1B} is only interesting if the bandwidth has to be reduced for another reason than for obtaining sufficient phase margin. A similar technique can be used to implement the outer compensation loop of nested-Miller compensation. Using Q_{18} the output voltage is supplied to C_{M2}. The signal current flowing through C_{M2} is therefore equal to the signal current flowing through C_{M2A} in the previous amplifier. Therefore, the signal current injected by C_{M2} into cascode Q_7 is just what is needed to implement the frequency compensation of the upper half of the amplifier. The signal current of C_{M2A} also flows through cascode Q_{18} and appears with the correct phase at the collector of Q_{18} for frequency compensation of the lower half of the amplifier. Again, the function of two capacitors is replaced by a single capacitor and further, the current mirror of the mirrored-nested-Miller topology is not needed. A drawback is that at low output voltages cascode Q_8 may saturate before output transistor Q_2 saturates. The result is an interruption of the frequency compensation yielding oscillations. Therefore, more design effort is required to investigate and solve this problem.

Using the first topology as shown in Fig. 5-57 a 200-MHz operational amplifier has been realized in a 0.8μm BiCMOS process. The complete circuit diagram is depicted in Fig. 5-59. The output transistors are protected using saturation limiter Q_{72} and current limitation resistors R_{13} and R_{14} which are inserted in the class-AB control circuit as described in chapter 3, section 3.8. To enhance the strength of the limiter, transistors Q_{75} and Q_{76} have been added. Using these transistors the value of current source M_{65} can be larger so that more current is available to control the intermediate stage. The frequency

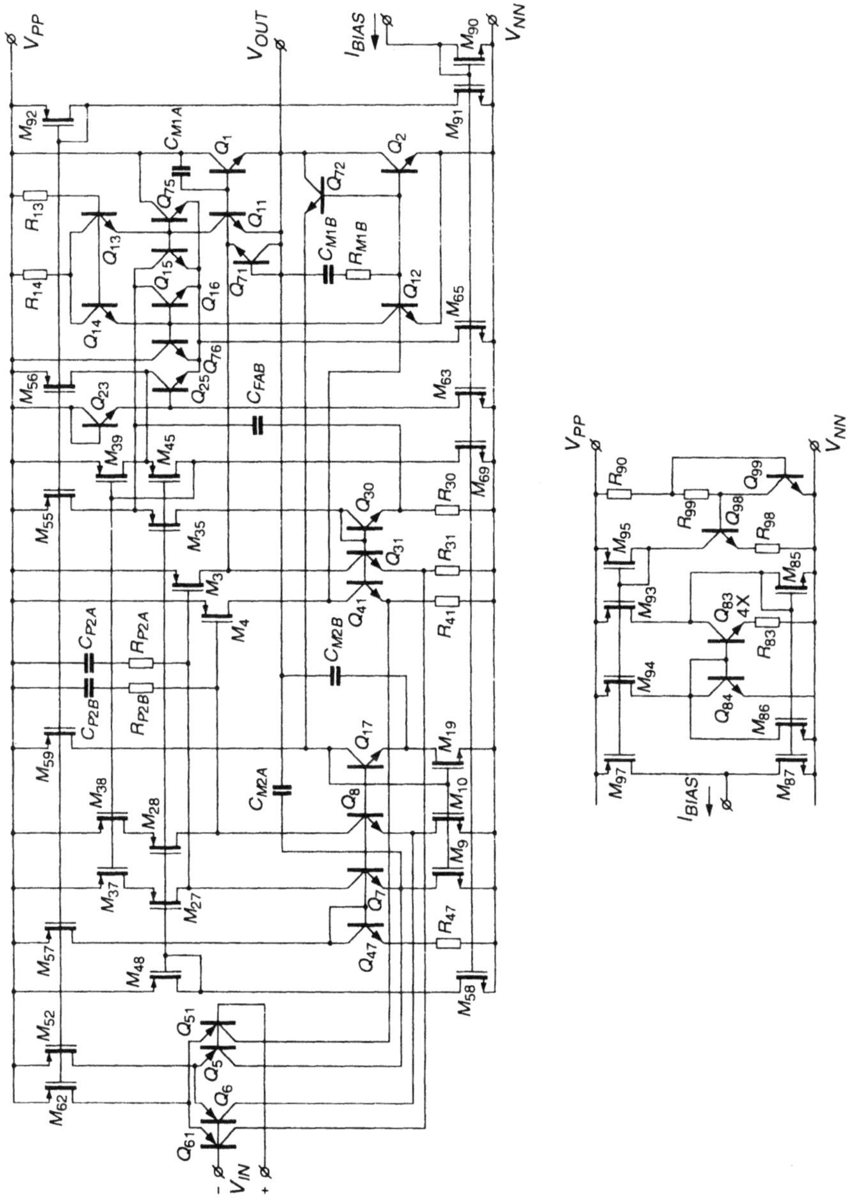

Fig. 5-59 *Complete circuit diagram of 200-MHz power-efficient BiCMOS operational amplifier.*

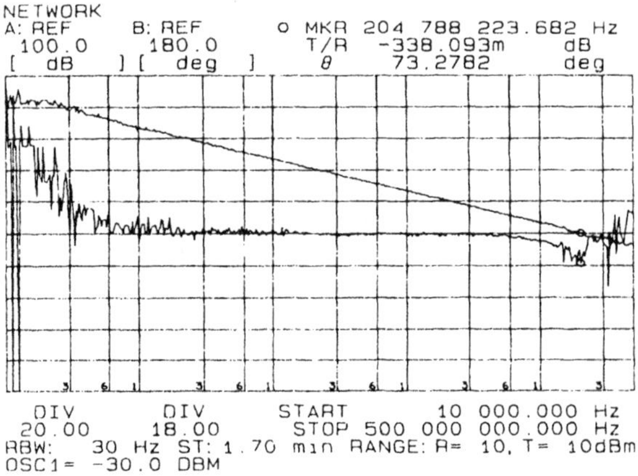

Fig. 5-60 *Frequency response of 200-MHz power-efficient BiCMOS operational amplifier.*

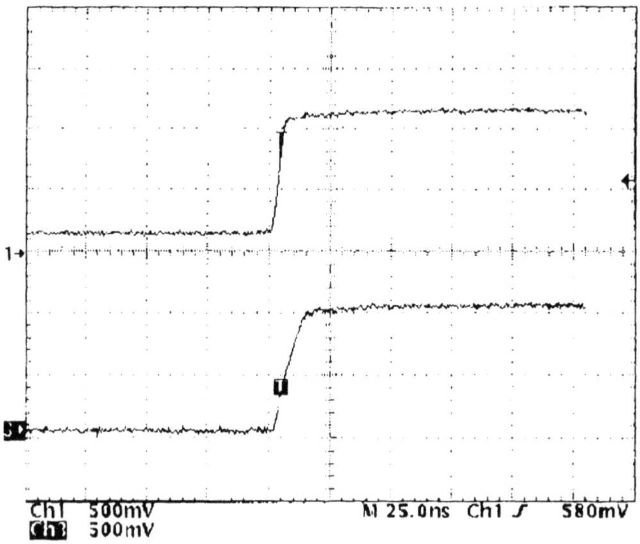

Fig. 5-61 *Response to a 1V step of 200-MHz power-efficient operational amplifier (upper trace shows input signal, lower trace shows the output signal; y-axis scale 500mV/div.).*

response is shown in Fig. 5-60 demonstrating a unity-gain frequency of 200 *MHz*. The response to a 1*V* step is depicted in Fig. 5-61. The amplifier settles to 1% within 15 *ns* and to 0.1% within 25 *ns*. The amplifier can operate on a supply voltage down to 2.5 *V*. Table5-8 summarizes the measurement results. A micrograph of the 200-*MHz* amplifier is depicted in Fig. 5-62.

Table 5-8 *Measured specifications of 200MHz power-efficient BiCMOS operational amplifier* (T_A=25 °C, R_L=10 kΩ, C_L=5 pF).

Parameter	value	Unit
Die area	0.05	mm^2
Temp. range	-55 .. 150	°C
Min. supply voltage	2.5	*V*
Supply current	2.9	*mA*
Peak output current	+2.5/-3	*mA*
Input offset voltage	0.5	*mV*
Input noise voltage (*f*=100 *kHz*)	4	$nV/\sqrt{Hz}$
CM input range	V_{NN}-0.3 .. V_{PP}-1	*V*
CM output-voltage range	V_{NN}+0.1 .. V_{PP}-1	*V*
CMRR	90	*dB*
Unity-gain frequency	200	*MHz*
Phase margin	70	°
dc-gain	90	*dB*
Slew rate	67	*V/µs*
Settling time (1%, 1*V*)	15	ns
Settling time (0.1%, 1*V*)	25	ns

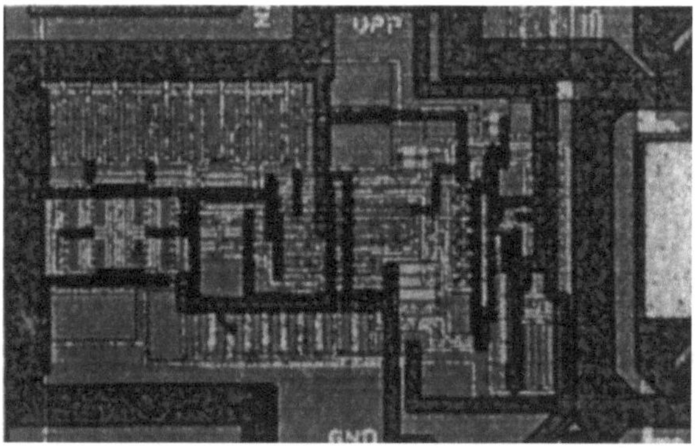

Fig. 5-62 *Photomicrograh of 200-MHz power-efficient operational amplifier.*

5.5 Conclusions

In this chapter complete amplifier topologies have been demonstrated in bipolar, CMOS, and BiCMOS technologies. We have seen how gain can be easily created in CMOS technology using cascoding techniques. Therefore, CMOS amplifiers can be implemented using a compact two-stage circuit. On the other hand, operating at the lowest supply voltage still requires at least three stages. In bipolar technology we can take advantage of the high transconductance yielding low noise and offset voltage, high bandwidth, and high output-current capability. But, usually more than three stages are required to obtain sufficient gain, and Darlington transistors are needed to obtain good power efficiency when driving large output currents. In BiCMOS technology the advantages of CMOS and bipolar technology can be combined, so that everything we want can be realized using a three-stage topology. Sufficient gain is created using CMOS cascoding techniques, while the combination of a PMOS driver with high input impedance followed by an NPN transistor with good output current capability is a very powerful tool to implement power-efficient stages that can drive large output currents. Finally, the multipath technique is a useful tool to combine the gain of a three-stage MOS circuit with the high bandwidth of a two-stage bipolar amplifier. An overview of amplifier topologies is given in Table5-9 where the following topologies are compared.

1. *Wideband Bipolar Topology (WBT).*
2. *Efficient Bipolar Topology (EBT).*
3. *Compact CMOS Topology (CCT)*
4. *Ultimate-Low-Voltage CMOS Topology (ULVCT).*

5. *Compact three-stage BiCMOS Topology (CBiT).*
6. *Low-Voltage BiCMOS Topology (LVBiT).*
7. *Wideband and Efficient BiCMOS Topology (WEBiT).*

The bipolar and BiCMOS topologies are the better choice if bandwidth and output-current capability are of prime importance. When high gain is also needed, BiCMOS should be applied. The compact CMOS technology requires little design effort and die area due to its simple structure, while the ultimate-low-voltage topology is necessary at the lowest operating voltages.

Table 5-9 *Comparison of overall amplifier topologies.*

Topology	WBT	EBT	CCT	ULVBT	CBiT	LVBiT	WEBiT
Bandwidth	++	o/+	o	o	o/+	o/+	++/+
Gain	o	+	+	++	++	++	++
Compact	o	-	++	+	+	+	+
Min. supply-voltage	-	o	+	++	o	+	-
Output-voltage range	-	-	++	++	++	++	-
Max. output current	o	++/+	-	-	+	++	+/++
Design simplicity	-	-	+	o	+	o	o

5.6 References

[1] J.H. Huijsing and F. Tol, "Monolithic Operational Amplifier Design with Improved HF Behavior", *IEEE J. Solid-State Circuits*, vol. 11, pp. 323-328, April 1976.
[2] J.H. Huijsing, R.G.H. Eschauzier, "Amplifier arrangement with multipath Miller zero cancellation", U.S. pat. appl. ser. no. 08/197,529, filed Feb. 10, 1994.
[3] J.E. Solomon, "The Monolithic Op Amp: A Tutorial Study", *IEEE J. Solid-State Circuits*, vol. SC-9, no. 6, pp. 314-332, Dec. 1974.
[4] E.H. Nordholt, "Design of high-performance negative-feedback amplifiers", Elsevier, New York, 1983.

[5] E. Seevinck, W. de Jager, P. Buitendijk, "A Low-Distortion Output Stage with improved stability for monolithic power amplifiers", *IEEE J. Solid-State Circuits*, Vol. SC-23, pp. 794-801, June 1988.
[6] J.H. Huijsing and M.J. Fonderie, "Multi-stage amplifier with capacitive nesting and multi-path forward feeding for frequency compensation", U.S Patent No. 5,155,447, Oct. 4, 1992.
[7] M.J. Fonderie and J.H. Huijsing, "Operational amplifier with 1-V rail-to-rail multipath-driven output stage, *IEEE J. Solid-State Circuits*, vol. SC-26, no. 12, pp. 1817-1824, Dec. 1991.
[8] R.G.H. Eschauzier, L.P.T. Kerklaan, and J.H. Huijsing, "A 100-MHz 100-dB Operational Amplifier with Multipath Nested Miller Compensation Structure", *IEEE J. Solid-State Circuits*, vol. SC-27, no. 12, pp. 1709-1717, Dec. 1992.
[9] K.J. de Langen, R.G.H. Eschauzier, and J.H. Huijsing, "A 1-GHz Class-AB Operational Amplifier with Multipath Nested Miller Compensation for 76-dB Gain", *IEEE J. Solid-State Circuits*, vol. SC-32, no. 4, pp. 488-498, April 1997.
[10] G.C.M. Meyer, "Thermal sensors based on transistors", *Sensors and Actuators*, vol. 10, pp. 103-125, 1986.
[11] W.C.M. Renirie, K.J. de Langen, J.H. Huijsing, "Parallel Feedforward Class-AB Control Circuits for Low-Voltage Bipolar Rail-to-Rail Output Stages of Operational Amplifiers", *Analog Integrated Circ. Signal Proc., vol. 8*, pp. 37-48, July 1995.
[12] K.J. de Langen, R. Hogervorst, J.H. Huijsing, "Translinear circuits in Low-Voltage Operational Amplifiers", in W.M.C. Sansen, J.H. Huijsing, R.J. van de Plassche: "Advances in Analog Circuit Design: MOST RF Circuits, Sigma-Delta Converters and Translinear Circuits", Kluwer Academic Publishers, Dordrecht, The Netherlands, 1996, pp. 357-38.
[13] R. Hogervorst, J.P. Tero, R.G.H. Eschauzier, J.H Huijsing, "A Compact Power-Efficient 3 V CMOS Rail-to-Rail Input/Output Operational Amplifier for VLSI Cell Libraries", *IEEE J. Solid-State Circuits*, vol. SC-29, no. 12, pp. 1505-1512, Dec. 1994.
[14] R. Hogervorst, J.H. Huijsing, *Design of Low-Voltage Low-Power Operational Amplifier Cells*. Dordrecht, The Netherlands: Kluwer Academic Publishers, 1996.
[15] W.-C.S. Wu, W.J. Helms, J.A. Kuhn, B.E. Byrkett, "Digital-Compatible High Performance Operational Amplifier with Rail-to-Rail Input and Output Ranges", *IEEE J. Solid-State Circuits*, vol. SC-29, no. 1, pp. 63-66, Jan. 1994.
[16] D.M. Monticelli, "A quad CMOS single-supply Opamp with rail-to-rail output swing", *IEEE J. of Solid-State Circuits*, vol. SC-21, pp. 1026-1034, Dec. 1986.
[17] E. Seevinck, "CMOS Translinear circuits", in W. Sansen, J.H. Huijsing, R.J. van de Plassche: "Advances in Analog Circuit Design: MOST RF Circuits, Sigma-Delta Converters and Translinear Circuits", Kluwer Academic Publishers, Dordrecht, The Netherlands, 1996, pp. 323-336.

[18] K.J. de Langen, J.H. Huijsing, "Compact 1.8V Low-Power CMOS Operational Amplifier Cells for VLSI", in *ISSCC Dig. Tech. Papers*, pp. 346-374, Feb. 1997.

[19] K.J. de Langen, J.H. Huijsing, "Compact Low-Voltage Power-Efficient Operational Amplifier Cells for VLSI", *IEEE J. Solid-State Circuits*, vol. SC-33, no. 10, pp. 1482-1496, Oct. 1997.

[20] R.G.H. Eschauzier, R. Hogervorst, J.H Huijsing, "A Programmable 1.5 V CMOS Class-AB Operational Amplifier with Hybrid Nested Miller Compensation for 120 dB Gain and 6 MHz UGF". *IEEE J. Solid-State Circuits*, vol. SC-29, no. 12, pp. 1497-1504, Dec. 1994.

[21] R.G.H. Eschauzier and J.H. Huijsing, "Multistage amplifier with hybrid nested Miller compensation", U.S. Patent Application, Serial No. 08/197,529, filed Feb. 10, 1994.

INDEX

R

S

T

U

V

W

MIX
Papier aus verantwortungsvollen Quellen
Paper from responsible sources
FSC® C105338

If you have any concerns about our products,
you can contact us on
ProductSafety@springernature.com

In case Publisher is established outside the EU,
the EU authorized representative is:
Springer Nature Customer Service Center GmbH
Europaplatz 3, 69115 Heidelberg, Germany

Printed by Libri Plureos GmbH
in Hamburg, Germany